G
186
753
992
Porter

W9-BYV-789

Laser Weapons

The Dawn of a
New Military Age

Laser Weapons

The Dawn of a New Military Age

Major General Bengt Anderberg
and
Dr. Myron L. Wolbarsht

Plenum Press • New York and London

Library of Congress Cataloging in Publication Data

Anderberg, Bengt.
    Laser weapons: the dawn of a new military age / Bengt Anderberg and Myron L.
Wolbarsht.
        p.   cm.
    Includes bibliographical references and index.
    ISBN 0-306-44329-5
    1. Lasers—Military applications. I. Wolbarsht, Myron. II. Title.
UG486.A53   1992
623.4′46—dc20

92-19697
CIP

ISBN 0-306-44329-5

© 1992 Bengt Anderberg and Myron L. Wolbarsht
Plenum Press is a division of Plenum Publishing Corporation
233 Spring Street, New York, N.Y. 10013

All rights reserved

No part of this book may be reproduced, stored in a retrieval system, or transmitted
in any form or by any means, electronic, mechanical, photocopying, microfilming,
recording, or otherwise, without written permission from the Publisher

Printed in the United States of America

To Professor Dr. Bjorn Tengroth,
a pioneer who continues to inspire all who work
with laser safety. He encouraged both of us in our
work and was always helpful.

Preface

The Persian Gulf war demonstrated to a huge television audience the effect of modern laser-assisted weapons. Everyone watching could see bombs and missiles scoring bull's-eyes when hitting and blowing up bridges, buildings, aircraft shelters, and various types of military equipment.

Almost every kind of laser device mentioned in this book, except laser weaponry, was used in large numbers during the conflict. Laser range finders of many kinds supplied vital information for soldiers and weapon systems. Laser target designators carried on aircraft, helicopters, and other vehicles, or hand-held by infantry soldiers, were used to direct Precision Guided Munitions (PGMs) such as the GBU-12 500-pound bomb and the Hellfire missile.

One of the main objectives during the air campaign besides hitting the enemy forces was to minimize civilian losses and damage to nonmilitary facilities. To accomplish this, the Coalition forces used PGMs. The attacks against populated areas such as Baghdad by the Stealth F-117 aircraft demonstrate the ability of lasers to assist in the destruction of military targets while minimizing civilian casualties.

Hellfire laser-guided missiles were used with devastating effect against tanks, vehicles, fortified bunkers, and radar sites.

During one night, eight Apache helicopters armed with Hellfires destroyed two radar sites deep in western Iraq. They fired 27 missiles in what was reported as the first hostile action of Operation Desert Storm. The success of this action opened a radar-free corridor that the aircraft of the Coalition could fly through unscathed to bomb the Baghdad area.

Approximately 20,000 PGMs were expended during the war. More than 60% of them were laser guided. The Persian Gulf war demonstrated the usefulness of different laser devices on the battlefield in assisting weapon systems, commanders, and soldiers. It must be left to the future, however, to find out how lasers may be used in their own capacity to destroy targets with the energy contained in their beams.

Acknowledgments

We thank Bengt Anderberg's wife, Margit, for her forbearance while he toiled on the initial version of the manuscript. We thank Ove Bring for educating us in the international law involved in a possible ban on the use of antipersonnel lasers as weapons and in war.

Many of our colleagues helped us form our opinions, but none should be held accountable for any misstatements made in this book or for the presentation of the details of any particular laser system.

This book is based entirely on public sources; see Appendix I.

Contents

xi

Introduction

Most battlefield weapons are extremely noisy and give off a flash as well as some smoke when fired, thus giving the soldier's position away and exposing him to immediate counterfire from the enemy. The ammunition from the weapons currently used by combat units has a curved trajectory with an appreciable time of flight; accordingly, adjustments must be made for this—aiming above the target to allow for the drop of the projectile and in front of moving targets. The path of the bullet (trajectory) is never flat; even very fast projectiles have some degree of curvature in their path to the target. All this makes it mandatory for the gunner to know the distance to his target more or less exactly as well as the speed of its movement. This is a fire control problem, and, in order to achieve the desired results, the effective number of grenades, shells, or bullets to be fired against each target often must be increased. Some of these fire control problems may be avoided by supplying the soldier with a missile with a guidance system. However, only high technology can give effective guidance to most missiles; this is very expensive and does not always work better than other solutions. The consumption of huge amounts of ammunition leads to logistical problems. Ammunition is very heavy and bulky, and there always is some risk of premature detonation when it is stored or transported. Against this back-

1

ground, it would clearly be advantageous if a different class of weapons could be fielded, specifically weapons that use ammunition with an extremely flat trajectory, are very speedy, and have only limited logistical problems. If such weapons are also invisible or less detectable when fired, as well as being silent, they would certainly also be more effective. At present, there is an option which seems to meet these requirements—the laser weapon. A laser beam has the speed of light, is absolutely straight, may contain a lot of energy, and, in most cases, is silent and invisible. Even cost-effectiveness is a possibility.

Laser technology is only 30 years old, but it is very diversified. There are already a variety of military applications, although there are many limitations restricting the use of lasers. Today, the armed forces in most countries routinely use a wide range of laser devices such as laser range finders and designators. In some countries, work is proceeding on more imaginative laser weapon concepts that will eventually fulfill realistic, yet very precise, military requirements.

The design of a specific laser weapon is heavily influenced by the characteristics of the intended target. If the desired effect of the weapon is to neutralize aircraft, helicopters, or missiles by burning holes through them or tanks by putting many miniature cracks (crazing) in the glass vision blocks to make them appear to be frosted, a very high energy laser has to be used with a power output on the order of several megawatts (MW). Such a laser would be a true antimatériel weapon. However, if the target is a sensitive electro-optical system or some other type of sensor system which has to be jammed or destroyed by a laser operating in a countermeasure mode, the choice will be a low-energy laser operating within the frequency bandwidth of the target sensor. This use of a laser can also be considered antimatériel. If the target is a soldier, there is one part of his body that is extremely sensitive to laser radiation—his eyes. It is sufficient to use a low-energy laser operating in the visible or near-infrared (near-IR) part of the spectrum to damage the soldier's eyes and, in effect, cause blindness. If the laser is to cause burn injuries to the soldier's skin or to set fire to his uniform, a high-energy laser is required. In either case, if the purpose of the laser is to blind or burn the soldier, it will obviously be antipersonnel.

THE HIGH-ENERGY LASER WEAPON (HELW)

The dominant effect of a high-energy laser weapon on the target is thermal. The transfer of energy from the laser to the target causes a succession of events: heating/melting and the evaporation of the target area depending on the reflectance of the target as well as its absorption and thermal characteristics. A very high energy level is required to cause this effect on a target several kilometers from the laser. The average beam power must be several megawatts during the required engagement time of up to a second or two.

High-energy laser weapons are now being studied mainly for antiaircraft and antimissile use. The advantages of using a laser against low-flying targets are obvious, as the warning and engagement times for such targets are very short, depending on the terrain and target speed. With conventional antiaircraft guns or missiles, there is always a risk that the enemy will penetrate the defense. The laser weapon, with its speed, straight line of sight, and short engagement time for each target, is expected to reduce or even eliminate such penetrations. Work has been going on in some countries for several years on different experimental high-energy laser weapons for general air defense and, in particular, for use by naval craft against sea-skimming missiles. If, in the end, any one system turns out to be successful, it could have the capability to engage five to ten targets up to a distance of ten kilometers away, within seconds of each other. The military advantages are obvious. Such a weapon would certainly be more effective than a conventional battery of guns or missiles. However, there are a number of disadvantages and complications. Any high-energy laser weapon will be bulky, costly, and noisy and will send out a large amount of heat in a characteristic fashion—a so-called thermal signature. This type of laser may consume comparatively large amounts of fuel. Another major problem is that the atmosphere will absorb a sizable fraction of the energy in the laser beam and be heated, which will make the laser beam wider and difficult to keep in focus on the target. To date, many attempts have been made to get around all these problems, and there seems to be some chance of success.

A high-energy laser may also be used against personnel, more or less in the same way as a very long range flamethrower. The beam could set fire to the soldier's uniform and other equipment and cause severe burns to the skin and eyes. However, the systematic use of very expensive high-energy laser weapons against personnel is not very cost-effective. The use of such high-energy systems will probably be limited to the protection of costly high-technology targets such as air bases, naval bases, high-level command posts, and aircraft carriers.

THE LOW-ENERGY LASER WEAPON (LELW)

A low-energy laser cannot burn through structures made from metal or other materials, even at the shortest distances on the battlefield. In spite of this limitation, low-energy laser systems may become even more useful as weapons than the high-energy ones.

Low-energy lasers designed for nonweapon use are already deployed by the thousands in armed forces. In range finding, the distance to the target is calculated from the time the reflected light takes to return to the emitting laser. This was one of the first military applications of laser technology, with field-deployed range finders from several countries being seen as early as the mid-1960s. Today, most armored fighting vehicles, armed helicopters, and ships and many aircraft have fire control systems which include a laser range finder or a laser target designator to guide a missile, smart bomb, or rocket to the target. An increasing number of artillery observers and soldiers associated with antitank weapons are equipped with small hand-held laser range finders.

Furthermore, low-energy lasers are widely used in weapon simulation systems for training, in communication systems, in laser gyros for guidance, in fusing devices, and in radar systems.

In the near future, low-energy laser weapons will become an increasingly important portion of the electro-optical devices developed and actually fielded. Low-energy laser systems will be used in two main modes: in an antimatérériel mode against sensors and in an antipersonnel mode against the human eye. There are many sensors on the battlefield that are sensitive to laser light. Antisensor laser weapons will be designed to destroy or blind image intensifiers, low-light televisions, and thermal sights—all essential for modern fire control systems. They may also be used to block or trick the sensors in missiles and smart munitions before these hostile projectiles can reach their target. However, the biggest impact that low-energy lasers may have on the battlefield in the future is the widespread use of anti-eye laser weapons.

THE EYE AS A LASER TARGET

The possibility of using a low-energy laser as an antipersonnel weapon is due to the high susceptibility of the eye to damage by a laser, since the eye's optics greatly magnify the brightness of the laser beam. This damage is even greater when any magnifying or light-collecting optical system is placed in front of the eye. Already, most of the existing military laser devices, such as range finders and target designators, are dangerous to the eye. The hazardous range for these devices is several miles, at least, and may be tens of miles if anyone in the target area is using binoculars.

An anti-eye laser weapon can have two main applications: temporary visual impairment such as flash blindness, discomfort, or veiling glare, or more permanent vision damage. As an illustration of the potential utility of an anti-eye laser weapon, a pilot who is flash blinded for 20 to 30 seconds or more during the final phase of an attack will most certainly have to eject from the aircraft or crash. Long-lasting damage inside the eye can occur through direct thermal damage (burns) or as a result of blood from a hemorrhage.

A laser weapon based on these principles will certainly have a dramatic impact on the probability of visually disabling an adversary, and it will certainly be a very effective weapon on the battlefield. An anti-eye low-energy laser weapon designed to blind enemy soldiers may very well be hand-held. It may be used alone or in combination with a rifle, machine gun, or any other weapon that uses a direct line of sight. Small hand-held anti-eye laser weapons would be comparatively cheap and certainly very cost-effective; the ammunition is stored energy in batteries. The batteries are certainly not expensive and would be readily available. Even the maintenance requirements of the laser system may be very small. However, these weapons will not only cause permanent injury and severe medical problems to large numbers of soldiers but will also have a marked psychological effect on the remaining personnel.

PROTECTION

One of the most important ways of blocking a laser beam from reaching the eyes or a sensor is by using a filter. However, the transmission of useful light through such filters is an important problem from a military standpoint. The ability to see adequately in a life-threatening situation is, of course, vital to the soldier. Visibility is already poor on the battlefield because of the smoke, haze, and terrain features and is even more limited at night, dawn, and dusk. Thus, the use of laser protective filters that further limit vision because of their low transmission of light is not a desirable feature militarily. A filter suitable for protection against frequency-agile laser weapons could make it more or less impossible for the soldier to see well enough to fight. At the moment, no safer alternatives for preventing eye damage exist that will not also hamper the soldier's effectiveness. Although other types of laser protection may become available, it is highly unlikely that any will be suitable for infantry use.

The impact of laser weapons on the future battlefield will be considerable. They will affect both operational methods and battle techniques. The use of an anti-eye laser as a common weapon on the battlefield will radically change the situation for the infantry-man, as there will be a new, silent, and less detectable threat for him to both use and guard against. In these respects, the antiper-sonnel laser weapon differs from other weapons. It will be neces-sary to include this type of laser threat in the training of most combat units. All armed forces will need to prepare new training manuals and design training devices based on new battle doc-trines adapted to the laser environment.

The question of medical treatment and medical resources is perhaps one of the most difficult and important issues to consider. The number of laser eye injuries will quickly rise to a level at which the present number of eye doctors and qualified surgical facilities will be totally inadequate. If large economic resources have not been directed toward this medical field in peace time, there will not be sufficient medical resources available for the injured sol-diers. Another very difficult problem is the extent to which the soldier's morale will be affected. The psychological impact on soldiers will be significant once they realize that observing the enemy may entail a significant risk of being blinded. If the medical resources are inadequate, the problem will become even worse, as few will risk losing their eyesight when adequate treatment is lacking. Furthermore, to these medical and psychological prob-lems, we may add the inevitable postwar consequences of laser weapons. After all wars, there have been many disabled soldiers, including some with severe visual impairment. This is a fact that any nation has to take into consideration. There is even some evidence that it is a more severe handicap to be disabled visually than to lose a limb or even to have severe multiple injuries. After any war in which anti-eye laser weapons have been utilized, the human and social costs of the resulting mass blindness will certainly be tremendous. This aspect of anti-eye laser weapons has already been the basis for questioning their military necessity on humanitarian grounds. Regardless of the final decision, the laser may change the face of war more than anyone ever imagined.

Desert Storm publicized a new sophisticated technology associated with modern warfare in a dramatic and almost unreal fashion. The public was shown the hardware that their high military budgets had purchased over many years. Many of these new weapon systems were supported by lasers. These lasers have many different roles to play. They are used to improve the accuracy of more conventional weapons, to find out what the enemy is doing, to defend against enemy weapons, and as components in communications systems. In this book, we have attempted to show the reader how lasers can fulfill these many seemingly contradictory functions and to explain the rationale behind each of the different uses, as well as their advantages and disadvantages. We have tried to project into the future as well as cover those lasers already tested on the battlefield. Some of these laser systems seem too fanciful to be real, as if some military procurement officer had mixed the sheets from a science fiction adventure in with his specifications for a weapons system to be put into production, but some such imaginative laser systems are already operational, and more will be in the future. However, it is our feeling that the laser which will have the greatest effect on the battlefield is a rather simple and quite unsophisticated one, namely, an antipersonnel laser weapon by which infantrymen can blind each other with an attachment to their rifles.

There have been vigorous debates in some countries about whether such weapons should not be deemed inhuman and their development ceased. The International Red Cross organized a number of meetings for the purpose of bringing together experts to evaluate the problems raised by the widespread use of anti-eye laser weapons in relation to international law. This is part of an international discussion now in progress on the subject of whether anti-eye laser weapons should be allowed. This subject is covered in detail in Chapter 8. Perhaps, this may lead to an international ban or some other type of restriction on the use of anti-eye lasers.

Some notes on terminology are appropriate here. In the military and scientific environment, the metric system is used in various forms (CGS, MKS, SI). Many quantities can be expressed as easily in the English (American) system (feet, yards, miles, pounds, and seconds), especially weapon ranges and weights. For these quantities, values in the English system will be given. However, the wavelength of light and other small dimensions are usually clearer when expressed in the metric system. A table of conversions is given in Appendix II.

ONE

Laser Technology

In this chapter, we will describe the physical principles that make lasers possible. Some laser applications will be mentioned briefly, but these descriptions will be expanded in later chapters. Laser technology is a very young science, only about 30 years old, but it has certainly had a very rapid development. A substantial part of the research and development in this field has been paid for by the military, and military requirements have, in many cases, guided the design of various lasers. However, the overall research and development (R&D) program has also resulted in broad-based civilian laser applications. Lasers are increasingly used in industry, medicine, and research and even are found in many homes. Perhaps the most common of all is the laser inside the mass-produced compact disc player, certainly numbering over 50 million.

There are many different types of lasers now in use, and even more will emerge from present R&D programs. All the various types of lasers have different properties; thus, can all of these many lasers do different things? Many already are used for various military applications, and more will be. To understand how lasers may be used for certain purposes and not for others, it is necessary to know something about basic laser technology. This is even more the case when it comes to laser weapons. It is not only

11

necessary to know why lasers do not vaporize people and pulverize buildings but also what lasers can do in today's military situations and what they may realistically be anticipated to do in the future with the expected advances in technology. Understanding current laser technology is a prerequisite for understanding the possibilities of laser weapons.

ORIGINS OF LASERS

Laser is an acronym for light amplification by stimulated emission of radiation. Stimulated emission occurs when an atom or molecule holds onto excess energy until it is stimulated by external energy, such as an impinging photon or electron, to emit its bound energy as light. This means that the light from a laser is different from the light that is normally seen in nature. Light from the sun or from light bulbs is emitted mostly spontaneously when atoms or molecules lose excess energy without any outside intervention.

The physical principles governing laser technology were explained as early as the beginning of the 20th century by the Danish physicist Niels Bohr. He showed in his quantum theory of 1911 that atoms lose energy (emit) only when electrons orbiting the nucleus move from a higher energy orbit to a lower one. They do not change energy in a stationary orbit. Albert Einstein suggested in a paper published in 1917 that an atom or molecule could be stimulated to emit light of a particular wavelength when outside light of that wavelength reached it. This was called stimulated emission. In 1928, R. Ladenburg proved that Einstein's theory was correct. However, it took many years before physicists discovered situations in which the stimulated emission was not overtaken by the spontaneous emission. The research work to make stimulated emission dominate in a given case was not accomplished until after the Second World War.

Physicists began looking for techniques to control how groups of atoms or molecules could be stimulated together to amplify light to much higher intensities. Charles Hard Townes of Columbia University was among the first to succeed. He worked in the early fifties on the problems of amplification of microwaves. Compared to light, microwaves have much longer wavelengths. In 1953, he completed the first maser, an acronym for microwave amplification by stimulated emission of radiation. Parallel work was performed in the USSR by Nikolai G. Basov and Aleksander M. Prochorov.

In 1957, Townes and Arthur Schawlow of the Bell Telephone Laboratories published a description of what they called an "optical maser." They described the conditions necessary to amplify stimulated emission of visible light. This meant an amplification of much shorter wavelengths than in the case of microwaves and was the final basis for the development of the laser. Gordon Gould, a graduate student at Columbia University, had a similar idea at about the same time, and many are still arguing about who deserves credit for the concept of the laser. Gould had to take his case to court and did not get the last of his patents issued in the U.S. Court of Patent Appeals until November 1987. However, Townes and the two Soviets, Basov and Prochorov, shared the 1964 Nobel Prize in physics for their work on the "maser/laser principle."

Townes and Schawlow's paper stimulated scientists in many quarters to try to construct the first working laser. In 1960, Theodore Maiman, a physicist working for Hughes Research Laboratories in Malibu, California, built and used the first laser. It was a ruby laser giving an extremely intense, short, red flash of light at a wavelength of 694 nanometers. Ruby is aluminum oxide with a trace of chromium oxide, which gives it a red color we call "ruby." The laser emission was pulsed. In the same year, a gas laser (helium–neon) generating a continuous beam of coherent light was developed at the Bell laboratories by D. R. Harriot, A. Javan, and W. R. Bennet.

Since 1960, the laser field has developed extremely rapidly. Today, in designing a laser for a particular purpose, it is possible to choose among many materials and methods. As a consequence, lasers differ greatly in power, wavelength, timing, and size. Lasers are now used for many different purposes in medicine, for mass machining of materials, for playing compact discs, for reading product codes in supermarkets, for measurement and inspection, in computer printers, and in a myriad of other civilian and military applications. This also means that there is a large and, in most cases, prosperous laser industry in many countries. This industry is, certainly, also the base for most military applications and a prerequisite for the worldwide proliferation of laser weapons.

From the very moment the first practical lasers appeared in 1960, their possible military uses were investigated. Since then, lasers have found ever increasing military applications. Lasers are widely used for range finding, for target designating, for guiding missiles (beam riding), and for many other purposes. So far, lasers have been used operationally in large numbers to support conventional weapons. Also, much R&D work has been done on antisensor, antisatellite, antiaircraft, and antimissile laser weapons. The possibility of designing and fielding antipersonnel laser weapons is also becoming a reality. As this book will show, there are many indications that anti-eye laser weapons are now well along the way to becoming operational on a worldwide basis. There is even good evidence that an anti-eye laser weapon may have already been used in field exercises and actual combat for some time in the British naval forces.

LIGHT

In the usual visual environment, many different colors are seen. This means that the human eye detects a variety of different wavelengths in the electromagnetic spectrum, which covers the range from extremely short gamma rays (below 3×10^{-11} meter) to very long, low-frequency waves (more than 3×10^4 meters) (Fig. 1.1). However, the human eye can only see a very narrow part of the electromagnetic spectrum. This should not suggest that there are actually sharp ends of the visible spectrum; there is rather a

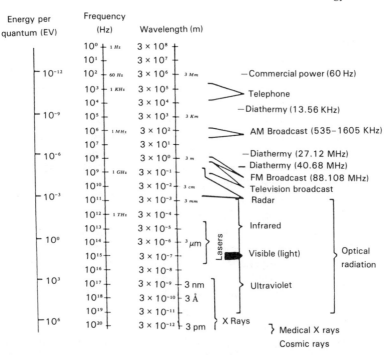

FIGURE 1.1. The electromagnetic spectrum. Optical radiation, including visible light, is a small part in the middle of the entire spectrum. The band covered by lasers extends roughly from 3 millimeters down through the ultraviolet and into the beginning of the X-ray region.

decreasing sensitivity at each end. Lasers, and thus laser weapons, emit radiation in that part of the spectrum corresponding to optical radiation (see Fig. 1.1), which includes the ultraviolet, visible, and infrared portions of the spectrum.

Electromagnetic radiation may be thought of as energy in transit. It is helpful to visualize energy in transit as extremely high velocity particles or packets of waves. In the case of optical energy, the particle is called a photon. The photon is also called a quantum of energy. The shorter the wavelength and, thus, the higher the frequency of the waves, the more energy is carried by the photon. However, light does not behave only as a stream of particles but also as waves. The electromagnetic wave motion is characterized by photon energy, wavelength, or frequency. Wavelength is the distance between the peaks of two consecutive waves or between any two identical parts of the wave. Frequency is the number of wave peaks passing a certain point per second. Frequency is measured in hertz (Hz). If a wave repeats itself at a rate of 10 times per second, its frequency is 10 Hz. The wavelength equals the speed of light divided by the frequency. The speed of light in a vacuum is a universal constant, 186,282 miles per second (3.3×10^8 yards per second).

Electromagnetic radiation is emitted whenever a charged particle loses energy in an electric field. In atoms, the charged particles—electrons—move in orbits around the nucleus. According to the quantum theory of Niels Bohr, atoms emit radiation in the form of photons only when electrons move from a higher energy orbit to a lower one. When energy is absorbed by the atom, the opposite transitions take place. When the electrons are in a stable orbit, they do not change energy.

In any particular atom, the electrons can occupy only specific and very well defined orbits. Thus, the single electron in a hydrogen atom may only occupy one of its six different orbits. The more distant the orbit from the nucleus, the more energy is stored in the atom. The lowest possible orbit in the atom is called the ground state. When an atom absorbs energy, if any electron gets more energy than is represented by the most distant orbit, it escapes from the atom completely—a process called ionization. The electrons fill in orbits from the innermost to the outermost orbit until there are as many electrons in orbit as protons in the nucleus. It must be remembered that under normal conditions there are more electrons in the ground state orbit than in any other level. Also, molecules have electronic energy levels due to their potential compression and expansion motions, which increase in complexity with the number of atoms and electrons in the molecule.

A transition of an electron to a lower energy level—that is, a lower orbit—releases energy at wavelengths typical of the specific substance. If energy is absorbed by the atom, the electron moves in the opposite direction, to a higher energy level. The light that we normally see in nature is the result of electrons dropping spontaneously and randomly to lower energy levels. This is the case even with man-made light sources such as incandescent and fluorescent lamps and television sets. This spontaneous emission is certainly not sufficient to cause laser action. For laser action, it is necessary to stimulate the emission by photons of exactly the right energy.

The simplest way to get a photon with the right energy, at the right place, and at the right time is to use the photon from an excited identical atom or molecule. However, to initiate a new transition, it is necessary that the receiving atom be in an excited state; otherwise, the photon would simply be absorbed and would not stimulate the emission of an identical photon. Thus, the atoms must be in an upper energy level. The greater the number of atoms in the upper energy level compared to that in the lower level, the greater will be the number of photons that can stimulate further emission than being absorbed. This is the inverse of the normal situation, in which more atoms are in the lower level than in the higher level. For this reason, the whole situation is called a population inversion.

There are two ways to accomplish a population inversion. One way is to enrich the higher orbits with electrons, and the other is to depopulate the lowest energy levels. If too many atoms or molecules are in a low level, this can end the population inversion and, thus, put an end to laser action. It is, therefore, necessary to both populate the upper level and depopulate the lower level at the same time if the laser is to operate continuously and not only with very short pulses. When a population inversion has been achieved, it must have a very "long" lifetime. In this context, long may be only a microsecond or millisecond, but this is much longer than is normal with spontaneous emission, in which case the lifetime of the excited state is on the order of nanoseconds. This long-lived excited state or level is called a metastable level.

Light and electricity are the most common ways to cause excitation to a higher level. When atoms absorb energy, they normally first reach a highly excited level and then spontaneously drop to the metastable level. When emission is stimulated from the metastable level to an empty lower energy orbit, the atom's energy level drops to a new and lower level above the ground state. This is a very simplified description of the whole process, but it will serve as a background for the following presentation of how different lasers work.

BASIC LASER FUNCTION AND LASER COMPONENTS

Only three basic components are really necessary for laser action: a lasing medium, a pumping system which supplies energy to the lasing medium, and a resonant optical cavity. Lenses, mirrors, shutters, saturable absorbers, and other accessories may be added to the system to obtain more power, shorter pulses, or special beam shapes. There are many solid, liquid, and gas materials which may serve as a lasing medium. The important thing is that the material must have atoms with at least one metastable level that lasts long enough for the population inversion to take place. Although laser action is possible with stimulated emission between two levels—a metastable energy level and the ground energy level—most lasers use more levels. Some lasers can simultaneously operate on many slightly different wavelengths, as the upper and lower energy levels may be divided into several sublevels. This is the case with the carbon dioxide (CO_2) laser, which is important in the high-energy laser weapons field.

The energy source may furnish energy to the laser—often called pumping—by optical, electrical, chemical, or nuclear methods. When the pumping system supplies energy to the lasing medium, the energy is stored in the form of electrons trapped in the metastable energy levels until a population inversion exists that is sufficient to enable laser action.

The pumping method that was originally used and is probably the least complicated is optical pumping. Theodore Maiman used it in the first laser ever built. Optical pumping is the use of photons for excitation. Optical pumping can use an electron flashtube filled with xenon gas or another laser or any other very strong source of light such as the sun. The lasing medium is exposed to light at the right wavelength to raise the atoms or molecules from the ground energy level to an excited energy level. Many lasers may be optically pumped with a flashtube, which emits a wide range of wavelengths. On the other hand, some lasers require pumping with photons in a very narrow range or band of energies, which can be accomplished by using another laser, generally of shorter wavelength than the one being pumped. Although flashtube pumping sources produce only pulses of laser action, continuous light sources can produce continuous-wave (CW) laser output.

Electron collision pumping (electrical pumping) is achieved by passing an electric current through the lasing medium. Lasers in which the lasing medium is a gas often use electrical pumping. When the electric current passes through the gas, it excites atoms and molecules, raising them to the excited energy level. Some gas lasers that use electrical pumping can produce beams that have CW laser output as long as the current passes through the gas. Electron collision pumping is also used to excite semiconductor action to produce laser action.

Chemical pumping is based on energy that is released in the making and breaking of chemical bonds, that is, in chemical reactions. Nuclear pumping means that the atoms and molecules capture the energy produced by nuclear reactions. Another quite different energy source is used in free-electron lasers, where a beam of electrons has energy taken from it as it passes through an array of magnets. All of these methods differ greatly from conventional optical and electrical pumping.

When population inversion is reached with the help of the pumping system, a few electrons decay spontaneously from the metastable level to a lower energy level and emit photons. A chain reaction starts when these photons hit other atoms or molecules and stimulate them to make the transition to lower energy levels, emitting photons of precisely the same wavelength, phase, and direction. To make this possible, the reaction must take place in a resonant optical cavity formed by mirrors that bounce the energy back and forth in a precisely regulated fashion. This helps amplify stimulated emission by reflecting some of the photons back into the lasing medium, which is achieved by using a mirror at each end of the cavity so that the beam passes through the lasing medium many times, and, thereby, the number of emitted photons is increased at each passage. To allow the laser beam to be transmitted out of the cavity, one of the mirrors is only partially reflecting and thus splits the light into one part reflected back into the cavity and another which is emitted as the laser beam.

The shape and arrangement of the mirrors in the optical cavity can differ among the various lasers. Two flat and parallel mirrors may seem to be a simple solution. However, these mirrors have to be parallel within a very high degree of precision in order to reflect the light rays back and forth between the mirrors without the unnecessary losses caused by even a slight misalignment. One way to avoid this problem is to use one or two curved mirrors. This method not only has the advantage of coping with misalignment, because the curvature focuses the light back to the other mirror, but even has the advantage of making use of light rays that are not exactly parallel to the axis of the cavity, as the curved mirror reflects most of the rays back into the cavity.

It should be noted that there are many solid, liquid, and gas materials which may serve as a lasing medium and many ways to provide energy to the lasing medium by optical, electrical, chemical, and nuclear methods. Furthermore, it is possible to arrange the resonant optical cavity in widely different ways. The possibility of choosing among these different materials and methods has led to the development of many types of lasers, and, as a consequence, laser beams may differ greatly in power, wavelength, detectability, and timing characteristics.

The selection of these materials and methods is critical when it comes to the design of laser weapons. The conditions within the resonant optical cavity affect the distribution of energy within the laser beam, which in turn determines how the laser beam will penetrate the atmosphere while traveling to its target. These factors will be discussed more fully in the following sections.

THE LASER BEAM

The laser beam has many unique qualities which can be manipulated in many ways by the use of different accessories that are added to the basic laser. The beam is characterized by its collimation, coherence, monochromaticity, speed, and intensity. Although beams from other light sources may possess some of these to a high degree, only the laser beam can possess all at the same time. This unique combination is the basis for many types of laser weapons and other military applications.

Collimation in a laser can be very high, which means that the radiation emitted by most lasers is confined to a very narrow beam which slowly diverges or fans out as the beam moves away from the laser source. The beam divergence is usually measured starting from the output end of the laser cavity. Calculating the beam size is a matter of elementary geometrical considerations. The beam divergence is normally a small enough angle so that the approximation holds that the sine and the tangent of the divergence angle have the same value, with the angle itself expressed in milliradians. The spreading out or divergence of most laser beams within battlefield distances is very small and is usually expressed as an angle from the source in milliradians (a milliradian divergence would mean that a beam would be 1 yard wide at 1,000 yards range, 2 yards wide at 2,000 yards, and so forth). Due to the wave nature of light, it is not possible to construct a laser that produces a perfectly collimated beam, that is, with no divergence. However, the divergence of a laser beam may be made much smaller than is possible with any other source of optical radiation now available.

If the laser designer wants this laser beam to be focused as much as possible on a small spot at long distances, the reciprocal relationship between divergence and the size of the output optics is used. When a beam with a very small divergence is required, large lenses must be used on the output of the laser. However, it is not possible to focus a laser beam to an exact point by using any real lenses. With ordinary lenses, the focal spot may not be smaller than a few times the wavelength of light. For most military purposes, this is certainly more than sufficient. In some high-energy laser weapon systems, a concave mirror is used to focus as much energy on the target as possible. As will be described in a later chapter, this is not without its own difficulties.

The laser is an excellent producer of coherent light compared to other light sources. The laser beam consists of an intense stream of electromagnetic waves, all of which have exactly the same frequency, phase, and direction of motion. This is due to the chain reaction whereby photons hit atoms or molecules, stimulating them to make the transition to lower energy levels, and the stimulated emission itself yields new photons of very nearly the same wavelength, phase, and direction. However, there are small variations in wavelength which, after a comparatively long distance, cause the temporal coherence to change significantly. The time over which the phase does not change is called the temporal coherence time. For laser beams, this may be kilometers or more. In different parts of the laser beam, the light waves may not have the same starting point, which disturbs the spatial coherence. The property of spatial coherence is essential to weapons applications, because a beam of light must be intense and well collimated in order to cause damage.

Laser light is often said to be monochromatic, that is, nearly of a single wavelength, and thus consists of a single color. However, there are usually small variations in wavelength (about a central one) which, after a comparatively long time, cause the temporal coherence or phase to change significantly. The time over which the phase does not change is called the temporal coherence time, and, if it is multiplied by the speed of light, the temporal coherence length is obtained. For laser beams, this may be kilometers or more. This slight variation is not important for any military application at present. Nevertheless, most lasers emit light over a range of wavelengths. It is up to the user to decide what operational requirements are necessary to determine exactly which laser wavelength among those possible is really needed. The designer of the specific laser may then, by using the appropriate laser optics, limit the emitted wavelengths to only one and, thus, make a monochromatic laser. It is possible to design lasers which emit laser energy at several wavelengths simultaneously. It is also possible to have a laser that is tunable continuously within a band, which could be important for some future laser weapon applications where any protective or countermeasures depend upon the laser wavelength being known beforehand with a high degree of accuracy.

The dye laser is one type of laser which may be tuned within a specific part of the optical radiation band. The tuning band for a specific dye may be found within the visible or the near-infrared part of the spectrum. There may be only a rather narrow tuning range for each specific dye used in the laser, typically, 50–100 nanometers. For example, one dye may allow tuning of the laser within the orange part of the spectrum, but obtaining a wavelength in another part of the spectrum requires changing the dye.

Another way to change the wavelength in a laser is to use the principle discovered by the physicist C. V. Raman. Although most of the light in a laser beam that is reflected from atoms or molecules is of unchanged frequency, a small part of the light has a different frequency. This fluctuation is created by combining energy such as the heat absorbed from the environment with the original energy from the pump. This change in wavelength is referred to as Raman shifting. In some laser applications, a Raman-shifted wavelength may be selected and emphasized. There are other laser applications that also allow for the selection or tuning of one wavelength among a variety of others that may be possible for a given laser medium.

The speed of the laser beam is enormous; the energy moves within the beam at the speed of light, which is roughly 180,000 miles (300 million meters) per second. For purposes of comparison, a supersonic missile has a speed of between 1,200 and 1,300 yards (1,100 and 1,200 meters) per second, while an attacking aircraft flies at 275–330 yards (250 to 300 meters) per second, and a helicopter at something like 100 yards per second. For this reason, the laser beam in a weapon may be considered to take a negligible time to arrive at the target or is spoken of as a zero-time-of-flight weapon. Speed is certainly one of the main advantages of the laser in weapons applications.

Lasers can operate in the continuous-wave (CW) mode or the pulsed mode. The mode of operation depends on whether the pump energy is CW or pulsed. A CW mode laser emits light steadily as long as it is turned on. A pulsed mode laser can have either one single pulse or repeated pulses, possibly on a regular basis in a train. The pulse repetition frequency (PRF) is the number of pulses a laser produces in a given time. The duration of the pulse (or pulse width) and the PRF may vary immensely between different lasers. Lasers are available with a PRF as high as several hundreds of thousands or millions of pulses per second. In a visible beam, such a pulsation will not be seen by the human eye, and the beam will appear to be CW.

One of the most important factors to a designer and user of laser weapons is the energy level delivered by the laser beam. Energy is the power emitted by a laser within a given time. The following equation can be used to calculate the intensity of the beam:

$$E = P \times t$$

where E is the energy in joules, P is power in watts, and t is time in seconds. The energy of repetitively pulsed lasers is calculated using the average power level emitted over a standard interval, which is usually one second. However, the energy level differs greatly among lasers. A high-energy laser weapon designed to down aircraft from several miles away may have several megawatts of power, while a low-energy helium–neon laser such as is used in a lecture hall pointer or a supermarket scanner usually has only a milliwatt or less of average CW power, although the CW power of a helium–neon laser can be as much as 50 milliwatts.

The highest level of power—the peak power—of each pulse can be considerably increased by rotating one end of a mirror or placing a shutter in front of it. The laser medium becomes excited, and a population inversion is produced which will not be depleted in a laser pulse until the mirror is unblocked. This means that the beam can be turned on for a few nanoseconds to a few microseconds. The total energy in every pulse is less than if the laser was operating with a longer pulse, but the energy is delivered in such a short time that the peak power of the pulse will be much higher than the laser could otherwise produce. This procedure is normally called Q-switching. Q-switched lasers can deliver very high peak powers—several megawatts or even gigawatts. A technique called mode locking produces even shorter pulses of a few nanoseconds. This means that a mode-locked laser will deliver a much higher peak power than a Q-switched laser with the same energy per pulse.

The laser beam is not homogeneous. The intensity drops off with distance from the center of the beam, and there are areas within the beam where the power level is much greater than the average across the beam. These areas are called hot spots, and the power level in these areas may, under some conditions, be 100 times higher than the average beam power level. Hot spots may be caused by inhomogeneities in the laser cavity, imperfections in the mirrors and lenses, or certain atmospheric conditions. The weather has an influence on the laser beam, and its effect is dependent upon the specific atmospheric conditions, laser properties, and energy level.

It should be kept in mind that the characteristics of a laser beam may vary enormously. This is very important when it comes to designing laser weapons of different types. For example, in an antipersonnel laser, the interaction of laser radiation with a biological system may make the pulsed-mode characteristics the more desirable choice. One way to change the characteristics of the laser beam is by changing the laser material.

TYPES OF LASERS

A laser is often classified according to the type of material used in its optical cavity. A solid-state laser uses crystalline or glass material, while a gas laser uses a pure gas or a mixture of gases. A semiconductor (diode) laser uses a specialized semiconductor material, and a liquid laser uses an organic or other type of dye in a liquid solution. Free-electron lasers (FEL) and X-ray lasers are other important laser designs. It is important to understand how these various lasers work, as the possibilities of designing and fielding laser weapons are tightly bound with the wide array of technology offered by different lasers. Those who want to have a closer look at the technical aspects of lasers may find it useful to read, for example, *Understanding Lasers* by Jeff Hecht published in 1988 by Howard W. Sams Company, Indianapolis, or *Safety with Lasers and Other Optical Sources* by D. H. Sliney and M. L. Wolbarsht, published in 1980 by Plenum Publishing Corporation, New York.

Solid-State Lasers

Solid-state lasers use a solid rod made up of the crystal or special glass that contains (or is "doped with") the atoms that absorb. The pump energy causes the population inversion and stimulates laser action. The characteristics of the different solid-state lasers depend on the active material used as well as on the substrate or host material. There are many suitable active materials such as the elements chromium, neodymium, erbium, holmium, cerium, cobalt, and titanium in a solid host material such as glass or an artificial crystal, for example, types of garnet or sapphires. In a ruby laser, chromium is the active material just as in the natural ruby gemstone. So far, neodymium is the active material that has found the most widespread applications. The solid-state laser almost always employs optical pumping by a flash lamp, arc lamp, or another laser. If the laser is pumped by a lamp, all the walls of the cavity are reflective and contain both the rod and the lamp. Q-switching is often used to shorten the pulses.

The very first laser was made from a synthetic ruby. The ruby was made by doping aluminum oxide with 0.01–0.5% chromium. The aluminum oxide crystal is the same material as the mineral clear sapphire, and the chromium atoms color it red or pink. Chemically, the ruby laser rod is the same as the mineral or gemstone ruby. The ruby laser emits a laser beam with a visible wavelength of 693.4 nanometers and produces a deep red light. The ruby laser is often Q-switched and produces short pulses of 15–20 nanoseconds duration with a pulse energy of something like 10–15 joules. The pulse repetition frequency is low, somewhat limiting the types of applications, but the laser may be made rather small and handy. Thus, ruby lasers found useful military applications as range finders in the early sixties and are still used for that purpose today.

The ruby rod has been replaced in many solid-state laser designs by neodymium atoms in a glass or crystalline material. Neodymium is the most common material used in solid-state lasers. Two of the most common lasers are Nd:YAG and Nd:glass lasers. The former is used more often and contains yttrium–aluminum garnet (YAG), which is a hard and brittle crystal, as the host material. Neodymium-doped glass is the second most used design. The wavelengths for the different neodymium lasers may vary slightly.

It is possible to use the neodymium laser for many military and civilian purposes by adding a variety of accessories. The wavelength and the pulse duration can be varied considerably. Certain types of interactions between the laser energy and various crystalline materials double the laser electromagnetic frequency. Doubling the frequency is the same as halving the wavelength, and thus the laser emission shifts from 1,064 nanometers (infrared) to 532 nanometers (green). The laser wavelength is thus moved from the near-infrared part of the spectrum to the visible. It is possible in this way to divide the wavelength also by three down to 354.7 nanometers and by four down to 266 nanometers, well within the ultraviolet part of the spectrum. These possibilities certainly make neodymium lasers usable for a tremendous variety of applications. Q-switching is often used in Nd:YAG lasers to achieve short pulses (10–20 nanoseconds) with high peak power (over 100 megawatts).

Neodymium lasers are widely used for military purposes as range finders and target designators. They are particularly well suited to these functions, since they can be made battery-powered, small, and handy. Furthermore, such devices can meet most rigorous military requirements such as reliability for outdoor use and resistance to temperature variations and other battlefield hardships.

Some solid-state lasers are tunable over a range of many wavelengths, as they emit laser energy whenever an excited atom drops from the upper level to a closely spaced group of lower levels above the ground state, usually called vibronic bands. Vibronic solid-state lasers are significant from a military stand-point, since their tunability is of critical importance when it comes to low-energy anti-eye laser weapons.

A good example of this type of laser is the alexandrite laser. Alexandrite lasers have been used mainly for research, with no widespread applications as yet. The basic host material is a synthetic form of a mineral known as alexandrite, and the active material added to the host is chromium. Alexandrite lasers can be tuned to wavelengths between 700 and 830 nanometers in the near-infrared part of the spectrum, and Raman shifting can efficiently convert this output to wavelengths within the visible part of the spectrum. This laser may be pulsed or operated continuously. The average power in a pulsed configuration may reach 100 watts. Some possible uses will be discussed in later chapters.

Another tunable solid-state laser is titanium in a sapphire host, termed a Ti–sapphire laser. The basic host material, aluminum oxide, is codoped with titanium and chromium. This laser can be operated in either the pulsed or the CW mode and is tunable from 660 nanometers in the visible up to 1,160 nanometers in the infrared. With frequency doubling, even shorter wavelengths can be generated. The output power has been reported to be around 15 watts.

A list of solid-state lasers and their typical operating wavelengths is given in Table 1.1. These are lasers commonly used in range finders and target designators.

Gas Lasers

Gas lasers use a pure gas or a mixture of gases to produce a beam, and there are many varieties on the market with very different properties. The emitted power can range from a thousandth of a watt up to millions of watts in pulses or CW form. The wavelengths produced by these lasers range from the ultraviolet—where an argon fluoride excimer laser emits at 193 nanometers—and continues through the visible and far into the infrared portion of the spectrum, where lasers can be found in the 30,000–1,000,000 nanometer region. The gas laser family includes a carbon dioxide laser (9,000–11,000 nanometers or 9–11 micrometers)

TABLE 1.1.   Solid-State Lasers

Name	Wavelength (nanometers)	Typical operation
Alexandrite	700–830	Pulsed/CW tunable
Erbium	850/1230/1540/1730/2900	Pulsed
Holmium: glass	1950	Pulsed
Neodymium	1064/1123/1318/1370	Pulsed
Quad	266	Pulsed
Trip	354.7	Pulsed
Doubled	532	Pulsed
Neodymium: glass	1060	Pulsed
Neodymium: YAG	1064.5	Pulsed/CW
Ruby	694.3	Pulsed
Quad	173.6	Pulsed
Trip	231.4	Pulsed
Titanium–sapphire	660–1060	Pulsed/CW tunable

which for many years has been one of the crucial high-energy lasers used in R&D for high-energy laser weapon applications.

The common gas laser is built around a tube which contains the gas. Mirrors are placed at each end of the tube. As in other lasers, one mirror is totally reflecting, while the other transmits slightly in order to allow the laser beam to leave the tube cavity. Most gas lasers use electron collision pumping, with an electric current passing through the gas. However, some gas lasers use optical pumping with flash lamps, and others use the energy generated by chemical reactions. A wide range of different pure gases and mixtures of gases are suitable for laser operation. It is only necessary that the gas have energy levels that are capable of achieving population inversion.

One of the best-known gas lasers on the market is the helium–neon (HeNe) laser. It produces a low-power, bright-red continuous beam with a wavelength of 632.8 nanometers. The power range is in the region of 0.1–50 milliwatts. Also, a HeNe laser emitting at a wavelength of 543.0 nanometers and with a power range of 0.1–1.0 milliwatts has recently become commercially available. It is also possible to design HeNe lasers with many other output wavelengths in the red and near-infrared portions of the spectrum. There are many reasons for the popularity of the HeNe lasers with a wavelength of 532 nanometer, the most usual one. They are simple, cheap, and can be cost-effective for many tasks. The lasers can work continuously for thousands of hours. Scanning HeNe lasers are used to read the standard International Product Code, which is increasingly used to mark packages in retail stores. Alignment and measurements at construction sites and in surveying are other common uses. Universities and schools train students in optical laboratories with HeNe lasers. The HeNe lasers do not have widespread military use but are found in ring laser gyroscopes, which are used to stabilize military aircraft and helicopters and to aid in the navigation of submarines. Such laser gyros may be made small enough to be suitable to control precision-guided munitions.

Argon (Ar) and krypton (Kr) lasers are very similar to HeNe

lasers. They have the same basic design, with electron collision pumping of the gas, but in this case the lasers use the rare gases argon and krypton and operate in the CW mode. Argon and krypton lasers differ from HeNe lasers in that they require active cooling and other modifications to give them higher power, up to 500 or 1,000 times more than that of HeNe lasers, and they are, therefore, quite a bit more expensive than HeNe lasers. The higher power possibility is one of their main advantages over HeNe lasers; another is emission at shorter wavelengths, which is necessary for some medical and industrial applications. Argon and krypton lasers allow for the selection of several wavelengths from the near-ultraviolet range through the visible region and down into the near-infrared part of the spectrum. They are used in medicine by ophthalmologists, mainly for the retina but also for glaucoma in other parts of the eye, and on the skin by dermatologists to remove birthmarks and tattoos. A common popular use of both argon and krypton lasers is to entertain audiences at light shows. From a laser weapons standpoint, argon lasers must be considered as possible low-energy anti-eye lasers which can blind temporarily with flash blindness or permanently by inducing laser lesions or hemorrhaging.

Helium–cadmium (HeCd) gas lasers use a vaporized metal as the lasing medium, and they can deliver a CW beam at power levels between 1.0 and 50.0 milliwatts. There are two possible wavelengths, 325 nanometers in the ultraviolet and 441.6 nanometers in the blue part of the visible spectrum. The design of the laser is similar to that of HeNe lasers except that it is necessary to heat the metal to produce a vapor. These lasers are not very expensive and have a considerably longer lifetime than argon lasers. There are other kinds of metal vapor lasers, especially copper and gold, which do not emit in the CW mode but only in the pulsed mode; these can have a high average power of between 1 and 50 watts. The copper vapor laser has two wavelengths, 511 nanometers in the green and 578 nanometers in the yellow, and the gold vapor laser has one at 628 nanometers in the red.

The earliest really high power laser was of the gas type, the carbon dioxide (CO_2) laser. The first one was built and demonstrated at Bell Telephone Laboratories by C. Kumar N. Patel in 1961. A CO_2 laser produces a beam in the infrared part of the spectrum at various wavelengths between 9,000 and 12,000 nanometers in either a CW or a pulsed mode. Although the CO_2 laser can operate on many of those wavelengths at the same time or be confined to a single wavelength, the most often used variety emits at a single wavelength of 10,600 nm. The efficiency of this CO_2 laser is extremely high compared to that of most other lasers, as high as 20%, with output powers from a tenth of a watt up to several megawatts. There are several possible types of CO_2 lasers with different properties, some of which are listed in Table 1.2. The more powerful CO_2 laser types are used for welding, cutting, and drilling metals and other materials. The high energy attained by CO_2 lasers has encouraged the military to try and develop

TABLE 1.2.   Different Types of CO_2 Lasers

Type	Design	Power	Type of output
Sealed-tube	A sealed tube; electric pumping	<100 W	CW
Waveguide	A small tube 1–2 mm across functions as a waveguide; gas flow	<50 W	CW
Longitudinal flow	A flow of fresh gas along the cavity; electric pumping	Several hundred watts	CW
Transverse flow	A flow of fresh gas transverse to the axis of the cavity; electric pumping	10,000 W	CW
Gas dynamic	See page 33	Megawatt	CW or pulsed
Transversely excited	A pulsed electric discharge is passed through the gas transversely		Pulsed

high-energy laser weapons. The gas dynamic CO_2 laser has been and still is the basis for many such military projects.

The CO_2 laser differs from most other gas lasers by burning fuel in oxygen or nitrous oxide instead of passing an electric current through a gas. This fuel may be a common hydrocarbon such as kerosene or methane, and the hot gas flows through a comb of nozzles, expands quickly, and achieves the population inversion required to amplify the energy. The gas then flows at supersonic speed through an optical resonator, where stimulated emission occurs, and the energy is emitted as a laser beam. The spent gas mixture is released through a diffuser into the atmosphere. The theory is rather simple, but putting it into practice may be complex. A compact turbine, a many-bladed fan, may supply the heated gas, while the spent gas carries off most of the stray heat at the same time. The overall function can be compared to that of a rocket motor in which the fuel and oxygen are forced into the combustion chamber under pressure and burned, and then the waste gases leave under a low pressure and absorb energy as they expand. In real life, there are a lot of difficulties involved in the design of a high-energy CO_2 laser weapon—the size and shape of the nozzles, the gas flow exit, and, not least, how to get the beam through the atmosphere focused directly on the target.

Another promising gas laser is the carbon monoxide (CO) laser. It operates in the wavelength region between 4,800 and 8,000 nanometers at about 70 different wavelengths and is still in the research phase. There are some difficult practical design problems to solve before a CO laser will reach the market. A CO laser would use a continuous beam with high power, and it would, in theory, be more efficient than a CO_2 laser in converting pump energy into laser power.

The excimer laser is an important new type of gas laser that was developed during the mid 1970s. The lasing medium consists of a mixture of noble gas (neon, argon, xenon, krypton, etc.) and a halogen (fluorine, chlorine, bromine, etc.). When the two gases are in the ground state, their atoms exist separately in the mixture.

When the atoms are excited to the upper level useful for lasers, molecules are formed, consisting of one atom from each gas. The laser is pumped with an electric current in short pulses. In general, an excimer laser beam has a wavelength in the ultraviolet and operates with short pulses, often less than a nanosecond in duration. The average power output may be over one hundred watts. Some typical excimer lasers are the argon fluoride excimer (ArF) at 193 nanometers, the krypton fluoride excimer (KrF) at 249 nanometers, and the xenon fluoride excimer (XeF) at 350 nanometers. Among other things, this new family of gas lasers is used for medical applications—ocular and vascular surgery—and in the electronics manufacturing process. These lasers find widespread military use in communication systems. One of their advantages is that they operate on a wavelength suitable for communications with submerged submarines. The excimer laser has also been a candidate within the Strategic Defense Initiative (SDI) program for use from the ground against targets in space with the help of relay mirrors positioned in space.

The chemical laser is similar to the CO_2 laser in that the laser action is fueled by the combustion of hydrogen with fluorine (HF) or deuterium with fluorine (DF). The vibrationally excited molecules emit photons, between 2,600 and 3,300 nanometers at more than 15 different wavelengths in the case of the HF laser and between 3,800 and 4,200 nanometers at about 25 wavelengths in the case of the DF laser. A chemical laser can operate in the CW mode, although the operational time is dependent upon how long it is possible to have gases flowing rapidly through the laser's cavity. Chemical lasers are similar to the CO_2 and other gas dynamic lasers and may also be compared with rocket engines. There is a great deal of military interest in chemical lasers, as it is possible to generate very high energy. In one concept called MIRACL (Mid-Infrared Advanced Chemical Laser), the U.S. Navy is said to have reached an output energy of 2.2 megawatts. However, in spite of reports that MIRACL has destroyed flying target drones, there are still huge practical problems to overcome.

There are some other gas types of lasers, such as the nitrogen laser and the far-infrared lasers, which have wavelengths longer than 20,000 nm. The nitrogen lasers are similar in type to the excimer lasers but have a more limited pulse energy. The far-infrared lasers seem to have very few practical applications.

A list of some of the more common gas lasers and their properties is given in Table 1.3. It may be noted that the most powerful lasers are in the infrared—to some extent, the further in the infrared, the more powerful the laser.

Semiconductor Lasers (Diode Lasers)

Transistors and solid-state (not gas or liquid) diodes are examples of conductors that are familiar to all because of their widespread use in television and radio sets. Under some conditions, certain types of semiconductor diodes can be used in lasers; when an electric current flows through the junction between the two materials forming the diode, the electrons are raised into an

TABLE 1.3.   Some Typical Gas Lasers and Their Usual Power Ranges

Name	Wavelength (nanometers)	Power (watts)	Typical operation
Helium–neon	543	0.0001–0.001	CW
	632.8	0.0001–0.05	CW
Krypton	350–647	0.001–6.0	CW
Argon	350–514.5	0.001–20.0	CW
Xenon fluoride excimer	351	0.5–30.0	CW
Argon fluoride excimer	193	0.5–30.0	Pulsed
Krypton fluoride excimer	249	7.0–100.0	Pulsed
Deuterium fluoride chemical	3,800–4,200	0.01–100.0 (Megawatts)	Pulsed or CW
Hydrogen fluoride chemical	2,600–3,000	0.01–150.0 (Megawatts)	Pulsed or CW
Carbon dioxide	9,000–12,000	0.1–15,000 (Megawatts)	Pulsed or CW

excited state that will emit light. The technical details of semiconductors are a vast scientific field in itself and are far beyond the scope of this book.

One of the differences between a semiconductor diode laser and the types of lasers already discussed in this chapter is that the beam of a diode laser is rectangular rather than round in cross section, and it diverges and expands very rapidly. That makes this laser beam more similar to an incoherent source such as a searchlight than to a laser, and other optical elements can be added to focus the beam according to the requirements of a particular application. A semiconductor laser offers a choice of wavelengths from 330 nanometers (zinc sulfide) up to 30,000 nanometers (lead salt). Table 1.4 presents an overview of semiconductor lasers.

Semiconductor lasers are used in huge numbers for compact disc audio players. They are also used in optical fiber systems such as telecommunications. New technological developments promise widespread availability of these lasers with an average power output from tens to several thousands of milliwatts and at lower prices. Semiconductor lasers may even be used in weapon guidance systems, in military communications, and in low-power antipersonnel (anti-eye) weapons.

The most common diode lasers are made of gallium arsenide. The basic model emits at a wavelength of 904 nanometers. It is possible to get shorter wavelengths, down to 750 nanometers, by using aluminum in the mixture. These lasers are of a very well known design; they are easy to use, and the low-power varieties are simple and cheap to manufacture. It is possible that gallium arsenide lasers in the near future will be efficiently frequency doubled and will thus generate wavelengths in the visible spectrum between 400 and 450 nanometers.

The indium gallium arsenide phosphide (InGaAsP) laser was developed to meet the requirements for longer wavelengths in conjunction with optical fiber technology. Such a laser may emit a wavelength somewhere between 1,000 and 1,700 nanometers. The most common versions have a wavelength of 1,300 nanometers.

A semiconductor laser that may replace the HeNe gas laser in

TABLE 1.4.   Some Typical Semiconductor Lasers

Name	Wavelength (nanometers)	Operation	Comments
GaInP	670–680		
GaAlAs	750–900	Pulsed	Best developed; highest power
Lead salt	2,700–30,000		

some applications such as reading postal zip codes or product bar codes at stores is the gallium indium phosphide laser, which emits in the visible region of the spectrum between 670 and 680 nanometers. It has a long lifetime, estimated at thousands of hours, and it may ultimately be cheaper and much more rugged than the HeNe gas laser.

Dye (Liquid) Lasers

One form of tunable laser uses an organic dye in a liquid solution as the lasing medium. A dye laser may also be pumped by another laser, for example, an argon, krypton, Nd:YAG, or nitrogen laser. A flash-lamp-pumped dye laser emits pulses with a duration from about a microsecond up to 500 microseconds. Laser-pumped dye lasers can be operated in either the pulsed or the CW mode, depending upon the mode of operation of the pumping laser. Each dye laser is tunable within a specific band, usually in the visible spectrum. If it is necessary to cover a broad part of the visible spectrum, then a combination of several dyes has to be used. A dye laser pumped by a pulsed nitrogen (N_2) laser may be tuned between 360 and 650 nanometers, while a CW argon pump laser allows tuning between 560 and 640 nanometers. The fundamental output wavelength of the dye laser may be frequency doubled or tripled as with other laser types. By using special techniques, a dye laser can emit ultrashort pulses down to the femtosecond range (10^{-15} second). Dye lasers are widely used in medicine and in research. The medical lasers may be used to treat birthmarks or eye problems, and also to shatter kidney stones. From a military point of view, it is necessary to discuss under what circumstances these tunable lasers in the visible spectrum are useful as anti-eye weapons.

Free-Electron Lasers

We can understand the free-electron laser (FEL) by comparing it to other electronic devices which share some common parts. The basic principle of an FEL is, to a large extent, similar to that used in a high-voltage linear particle accelerator, in which a beam of electrons is accelerated to a very high speed by an electric field. These electrons are alone and free of any atoms. When the electrons have been sufficiently accelerated, they are passed through a magnetic field produced by placing a series of magnets in a row, with every other magnet of reversed polarity. This magnetic field causes the electron beam and the electrons in it to wiggle or change direction sharply. As the electrons change direction, they emit and absorb energy. If the design of the laser is proper, the electrons will emit more energy than they absorb. The electrons, after changing directions but before emitting energy, can be thought of, for comparison purposes, as the population-inverted energy levels of a conventional laser, and, by stimulated emission, the "excited" electrons will emit coherent radiation. This energy output is formed into a laser beam in a conventional way by the use of mirrors in a resonator cavity.

FEL technology is still very much in a research phase. If all practical difficulties can be solved, the main advantages will be extremely high efficiency, tunability over a very broad part of the spectrum, and high intensity. M. Foley, a researcher at Los Alamos National Laboratory, estimates the efficiency of a highly developed FEL as 20–24%, which is really a very high efficiency compared to that of other tunable lasers. In theory, the wavelength range extends all the way from the microwave region to the ultraviolet, but any particular FEL is limited to only a specific part of that range. However, even so, this range is much broader than what is available with any other tunable laser. The wavelength may be tuned by changing either the magnet spacing or the speed of the electron beam. However, the property that has triggered the large funding for R&D of these lasers is the possibility of generating extremely high powers which could be used in laser weapons within the SDI program.

Many difficulties and practical problems have to be solved before FELs can be considered completely successful. So far, ultrahigh power levels have not been reached in any of the desirable wavelength regions. However, it seems likely that an FEL will be able to emit an average beam power of several million watts. The pulse length of an FEL may be very short, in the picosecond (10^{-12} seconds) region. As yet, existing FELs are quite big and heavy. This means that an FEL weapon cannot be placed in space but has to be stationary on the ground; therefore, mirrors in space must redirect the beams onto the incoming target. Current programs are aimed at designing a much smaller and more compact FEL that could be tuned between about 1,000 and 10,000 nanometers.

FEL technology is far from mature, and it will certainly take some years before a useful FEL is fielded. However, the FEL's promise of high power, efficiency, and tunability will certainly put pressure on those developing FEL technology to make it work.

Miscellaneous Lasers

There are some very specific laser projects whose goal is to produce very short wavelengths in the X-ray region. One such project is the development of an X-ray laser pumped by a nuclear explosion or eventually with an extremely powerful laser operating with very short pulses. Another possibility, at an even shorter wavelength, is the gamma-ray laser. These efforts have been made within the SDI program in response to the military's requirement for a space-based laser with the ability to kill missiles thousands of kilometers away. A tremendous number of practical problems need to be solved before any of these concepts can be considered even a limited success. Furthermore, these lasers require pumping by a nuclear explosion or by a giant laser. At this time, none of the designs can be considered suitable for battlefield laser weapons. However, a lot of money has been relegated to such laser projects for "space war," and certainly more will be in the future.

CIVILIAN LASER APPLICATIONS

Lasers are being used to an ever increasing extent in everyday civilian life. The reason for this is the availability of suitable and cost-effective lasers for many applications. The opportunity to choose a wavelength, mode of operation, and power requirement all suited to a very specific task is now a reality. Below, some examples are presented that illustrate how frequently lasers are now used outside of the military. The laser industry is a massive, worldwide phenomenon that is steadily growing. All highly industrialized countries have some R&D programs, in addition to their own well-developed laser manufacturing facilities, while almost all other countries have the beginnings of some kind of laser industry.

Lasers are used in medicine for many delicate tasks. The use of lasers for the treatment of various diseases in the eye is perhaps one of the more commonly known applications. Ophthalmologists may repair a detached retina, photocoagulate large portions of the retina to control diabetic vascular pathology, seal leaking blood vessels, photocoagulate the eye to lower the intraocular pressure in glaucoma, or puncture a cloudy membrane. In the future, perhaps, the cornea may be sculptured to correct refractive errors with the help of a laser. This could eliminate the need for most spectacles and contact lenses. For these different treatments, CO_2, argon, neodymium, dye, and excimer lasers can be used. Surgeons may use a laser as a surgical knife, removing tissue with, for example, a CO_2 laser beam. Birthmarks may be removed by dermatologists using an argon or dye laser. Kidney stones can be shattered by a pulsed dye laser, with the energy carried through an optical fiber to the kidney stone. Other medical uses are destruction of some cancer cells and treatment of bleeding ulcers in the stomach as well as in various types of diagnostic work. This is only a brief sample of a myriad of medical applications for lasers. There will certainly be even more as new applications constantly emerge in the field of laser treatments.

Lasers are also used in construction for alignment, leveling, and surveying. Alignment lasers may be used in mining, tunnel construction, pipe installation, and road work. The lasers used for this purpose are normally HeNe lasers with a visible red beam, but semiconductor diode lasers may supplant most of these.

Lasers can be used industrially for precision drilling, marking, and welding and cutting of many metals, plastics, and even diamonds. The properties of the specific laser that should be used in each case are dependent upon the task and the material that is being worked with. The effect of the laser is usually thermal, which means that it is necessary to use a laser with a wavelength that is heavily absorbed by the target material. For a given material, the differences in absorption between various wavelengths are often quite large. As an example, for copper as the target material, the argon laser is absorbed 56% at 500 nanometers, the ruby laser 17% at 694 nanometers, the Nd:YAG laser 10% at 1,064 nanometers, and the CO_2 laser 1.5% at 10,600 nanometers.

Lasers are also used in the printing industry, in the manufacture of electronics, in reading information on reflective compact discs and video discs, in optical fiber communication, in optical computers, in laser displays and laser art, and in supermarkets to read the International Standard Product Code on packages.

SUMMARY

Present-day laser technology is very extensive and diversified, and, within certain limits, it allows for many civilian as well as military applications. Military staff, defense research institutes, and defense industries are constantly looking for new laser concepts that are suitable for military application and that will fulfill the very tough but realistic battlefield requirements. Many new military laser systems will most certainly be designed and fielded, and most countries already have a laser industry to back up their military needs. Thus, if and when realistic battlefield laser weapons concepts pass through the research and development phase, there will be a strong laser industry already in existence to mass-produce these weapons.

MILITARY LASER RANGE FINDERS

Range finding was the first military application of the new laser technology. Operational range finders were introduced into the armed forces as early as the mid-sixties, only five years after Theodore Maiman presented the first working laser. Many of these early field-model laser range finders are still in use today. Since then, thousands and thousands of laser range finders have been delivered to the defense forces in many countries all over the world. Today, laser range finders are a necessary part of most modern fire control systems, and a hand-held version is an indispensable aid to the modern infantry.

The availability of accurate and quick information on what the distance is between the target and weapon is essential to achieve a high hit probability with direct-line-of-sight weapons including tank and antitank guns, antiaircraft guns, and many weapons and guns carried on airplanes and ships. The probability of hitting the target with the first round fired is substantially higher if the firing distance is known accurately rather than if the gun crew has to act on information that contains some degree of error. Of course, the laser is not the only range-finding method available. Before the laser, optical type range finders were extensively used, but they are simply not as reliable or as accurate as laser range finders. Optical range finders are also difficult to operate, slow, and rather expensive compared to laser range finders. Radar is another possible method but again is more expensive and certainly bulkier and somewhat less accurate than a laser range finder. Laser range finders are small, highly accurate, low in weight, and, at least in some versions, comparatively cheap. For these reasons, they are completely replacing the older optical range finders. Lasers are used for range finding in all military service branches, on land, at sea, and in the air—even replacing radar in many cases.

The principle behind laser range finders is rather simple. A very short laser pulse—about 10 to 30 nanoseconds—is emitted toward the target, and the time is measured until a reflected signal is returned. As the speed of the laser pulse is known precisely, it is possible to calculate the distance to the target with great accuracy. When the trigger of the laser is pulled, a storage capacitor provides energy to the laser and also gives a reference pulse of light to be used as a start signal by the high-speed calculator. The laser beam is reflected from the target, and a tiny fraction of that reflected energy reaches the collecting optics (a telescope) and is focused on a laser detector. The resulting signal from the detector is amplified and compared to a reference standard to determine whether or not the signal is strong enough to be the "true" echo. Otherwise, the ambient light that reaches the detector could trigger a false signal. The calculator gives the time between outgoing and incoming laser light by counting the pulses from a quartz-controlled clock. The distance calculated by this procedure is usually within one percent of the true value.

Almost all laser range finders on the battlefield today are based on solid-state laser technology. A flash-lamp-pumped solid-state laser generates a train of pulses, each lasting several microseconds. This type of output is almost useless for range finding, and the lasers therefore have to be Q-switched to get even shorter pulses—approximately 10 nanoseconds long. The first laser range finder used a single-pulsed ruby laser operating at a wavelength of 694.3 nanometers. Although ruby lasers, both single and repetitively pulsed, are still used fairly often in tanks and in other weapon systems, the most common range finder in use today employs an Nd:YAG or Nd:glass laser with an output wavelength of 1,064 nanometers.

One of the disadvantages of both the ruby and the Nd:YAG range finder is the hazard to the human eye. This danger will be described in detail in Chapter 6 in the context of possible anti-eye laser weapons. The eye hazard makes it difficult or impossible to use ruby and neodymium lasers. These lasers can only be used on closed rifle or artillery ranges or within military training areas using very strict range-type safety regulations similar to those used in the firing of direct-line-of-sight weapons such as rifles, machine guns, and antitank guns. This limitation has resulted in the R&D of several varieties of laser range finders that are safe to the eye for use where range-type safety regulations cannot be enforced. One example is the CO_2 range finder at 10,600 nanometers, which, at power levels adequate for range finding, is perfectly safe to the eye. The CO_2 laser range finder also has other advantages such as better penetration through smoke and better spectral compatibility with the increasingly employed infrared sensors. Special amplifiers for the detector signal extend the range but are expensive options. The major drawbacks to the CO_2 laser range finders are their high cost and limited range for ground-to-ground use due to the reduced transmission through the moist air layer next to the ground. The high cost is due not only to the high procurement price but also to their complexity, which raises maintenance costs sharply. The CO_2 laser will not completely replace the Nd:YAG laser. Still, CO_2 laser range finders will probably be fielded, for example, in tanks as an integrated part of the fire control equipment, which also contains a thermal sight that can use the CO_2 laser to illuminate or mark the target.

Other options for eye-safe lasers in addition to the CO_2 laser are the erbium:YAG laser, a solid-state laser at 1,540 nanometers, and a Raman-shifted Nd:YAG laser whose wavelength is changed by a pressurized gas cell from 1,064 to 1,540 nanometers. The latter wavelength is in a region safe for the human eye in low-energy applications. Another alternative is a semiconductor laser emitting between 1,500 and 1,600 nanometers. None of these alternatives is as efficient as an Nd:YAG laser, but when the amount of energy used to supply the range finder is of little or no significance, erbium or similar laser types may be used. However, for battlefield tasks that require a laser range finder with high pulse repetition frequency, there is really no alternative to the Nd:YAG laser at present.

Laser range finders made today are used separately or are integrated into larger, more sophisticated fire control equipment. Small hand-held laser range finders, the size and shape of an ordinary pair of binoculars, are now in use. An excellent example of such a laser is the British Laser Gage LP 7. It is an Nd:YAG laser which can measure up to 6 miles with a resolution of 5 yards. It is battery driven and is said to be able to fire at least 600 ranging shots without a battery change. This laser range finder is being used in many countries in increasing numbers by infantry forward scouts and forward artillery controllers. Fire control teams that direct mortar and artillery action, antitank weapon crews, and many others on the battlefield need to know the distance to various targets as well as important terrain features. The LP 7 is proof that it is possible to mass-produce laser equipment which fulfills rigid military requirements. Other hand-held range finders similar in function and quality to the LP 7 are also available such as the U.S. AN/GVS5 and the Yugoslavian RLD.

Even bigger lasers that have more power than the hand-held models, yet still can be classified as portable, may be used on a vehicle or placed on the ground on a tripod. One example of such a laser is the Soviet KTD-1 Nd:YAG laser range finder. This range finder is reported to be capable of measuring distances between 100 and 10,000 meters depending on the target. With a target area of 0.2 square yards, the range will be nearly $1\frac{1}{4}$ miles. For a target the size of a tank, the range will be something like 5 miles, and a building will be reliably ranged out to about 10 miles. Furthermore, this particular range finder is reported as being capable of working in temperatures ranging from $-40°$ to 145°F and from sea level up to $2\frac{1}{2}$ miles. Its laser is said to have a lifetime of 100,000 measurements. Similar laser range finders are produced in the West, and these are mostly used by artillery control teams.

Modern fire control and surveillance systems often combine laser range finders with direct sighting systems such as television cameras or night vision equipment. This means that most armored fighting vehicles and helicopters, warships, and aircraft will have systems which include a laser range finder. Swedish tanks, similar to many other tanks, have Nd:YAG range finders. Older tanks such as the U.S. M60 have ruby laser range finders, and the latest tank designs incorporate CO_2 laser range finders. The U.K. Royal Air Force has a Laser Ranger and Marked Target Seeker (LRMTS) in service on the Harrier and Jaguar attack aircraft. The LRMTS includes a Nd:YAG laser range finder which offers the pilot a maximum measuring distance of $5\frac{1}{2}$ miles. This is just one of many integrated fire control systems for aircraft that also includes a laser range finder.

LASER TRACKERS

Most laser range finders operate with a single-shot technique—the operator finds the distance to the target by aiming the range finder, pressing the trigger, and watching as the measured distance appears on his display. This is fine for a stationary or slowly moving target such as a tank, but it is not satisfactory when it comes to a rapidly moving enemy aircraft, missile, or helicopter. It is necessary to track such targets and to get nearly continuous information on the distance to the target as it quickly changes. To get real-time information on the engagement range, it is necessary to update the information at least 10 to 20 times a second. This problem can be solved by using a high-repetition-rate laser range finder. High-repetition-rate range finders pose some design problems as compared to the single-shot range finders. The rather simple single-shot device has to make comparatively few shots during its entire lifetime, while the high-repetition-rate laser range finder will make millions. For example, a 20 pulse per second high-repetition-rate range finder in one hour of continuous use will emit 72,000 pulses, as opposed to about 100 pulses for a single-shot range finder. The design of such range finders is, therefore, much more sophisticated. Also, the high-repetition-rate laser needs a longer range in order to follow fast-moving enemy targets at distances of more than six miles. Thus, if the laser range finder is to have the same reliability as the rest of the fire control system, its expense will be much higher than that of the single-shot hand-held device.

The high-repetition-rate laser is used in antiaircraft and antimissile systems, especially to track and shoot down low-flying aircraft and sea-skimming missiles. This has led to the design of a laser range finder that can help in tracking this special type of target. The laser detector is split into four quadrants to give the strength of the incoming signal from the target relative to the laser beam. The angular direction information is used to direct the platform upon which the fire control equipment is placed. The tracking accuracy is better than 0.3 milliradians at ranges of up to $3\frac{1}{2}$ miles. It is highly probable that the laser missile tracking technique will be developed further, allowing longer engagement distances for very small and fast targets. The present trackers have, in bad weather at least, a rather limited maximum range for tracking these small and elusive targets.

LASER TARGET DESIGNATORS

There has always been a military need for weapons capable of pinpoint accuracy. This is especially true when the target is small and well defended. Until now, there have only been two alternative ways to deal with this kind of military situation—either get close enough to the target to make certain of a hit or use some kind of blanket bombing or shelling over a fairly large area. Closing in on the target may be extremely dangerous if it is well defended and could lead to a high casualty rate. On the other hand, bombing or shelling may not be effective in destroying the target or may require excessive amounts of ammunition. This has forced the military into the development of smart bombs or shells which can easily pinpoint specific targets. The laser target designator is an important part of most of these sophisticated munitions systems.

In smart munitions systems, a coded laser beam from the laser target designator is directed at the target. The reflected pulses from the target are scattered in many directions. They are detected by the missile's laser target seeker, which is a sensor on the head of the missile that responds to the same code as in the beam. The missile, which normally is fired from a completely different place, will thus home in on the target and destroy it. The missile or bomb does not follow the beam from the designator to the target but rather sees a reflection from the target; that is, the system is not a beam riding type. The coding of the laser beam is of vital importance. The only equipment that should be able to read the code is the target seeker in the actual munition. Otherwise, it would be possible for the enemy to steer the missile away from the target with its own laser beam. The coded beam also has the advantage of allowing simultaneous marking or designation of several different targets in the same area, each with its own designated missile and each with a different code to prevent interference between them.

In most cases, the laser used in the target designators is a pulsed Nd:YAG laser with a very high pulse repetition frequency. The CO_2 laser is an alternative to the Nd:YAG laser, especially when penetration through smoke and haze is taken into consideration. However, the CO_2 laser system is relatively expensive, not only because of the cost of the laser but also because of the cost of the laser detector in the missile, which means that the Nd:YAG laser designator will be the favored choice for a long time.

The laser target designator may be mounted on a tripod to be portable for use by infantry, as in the case of the British lightweight ground designator from GEC Ferranti Defense Systems (Fig. 2.1), or it may be integrated into a more complex stationary fire control system. Many weapons are already fielded in army, navy, and air force applications and are guided by laser target designators. The U.S. Army uses an antitank missile called Hellfire and an artillery-delivered antitank shell called Copperhead. The U.S. Air Force has several types of laser-guided bombs (LGB) and missiles as does the U.S. Navy. Many other countries have also developed these types of weapons—for example, the French have the Laser Guided Bomb (BGL) family.

A vast amount of money has been spent on the R&D of airborne laser target designators. The first version fielded was PAVE LIGHT, designed in response to the difficulties that U.S. aircraft had in hitting targets in North Vietnam from the air. This was the first target designator. It was operated visually from the back seat of a Phantom aircraft which had to circle the target and keep the

FIGURE 2.1. British lightweight ground designator FERRANTI INTERNATIONAL, type 306. Photograph courtesy of GEC Ferranti Defense Systems, Limited.

beam directed onto the target while other aircraft dropped their laser-guided bombs. Development proceeded rapidly to a target designator contained within a pod mounted on the wing combined with a television tracker and a laser range finder. Later versions also operated at night, giving a 24-hour capability. The latest U.S. system is called LANTIRN and consists of a navigational pod and a targeting pod which have a modern laser target designator/ranger and an automatic tracker, with a Forward Looking Infrared (FLIR) imaging system. Automatic tracking and laser target designation allow the aircrew to concentrate on countering threats and getting the aircraft out of the danger area. Meanwhile, the tracker and stabilization system keep the laser beam on the target accurately to ensure a high hit probability.

There are similar airborne target designator systems in other countries. A Thermal Imaging Airborne Laser Designator (TIALD) is placed in the British Tornado aircraft. It offers a day/night capability for target acquisition with identification and laser designation for laser-guided bombs. In France, the Convertible Laser Designator Pod (CLDP) is currently under development. It has been tested together with an AS.30 L laser-guided missile in darkness at a distance of 5 miles from the launch to the target. The aircraft was at a height of less than 300 yards and had a speed of 560 miles per hour. Most modern systems allow operation by the pilot in a single-crew aircraft. There can be no doubt that a similar development of airborne laser target designators is going on in the remnants of the Soviet Union and other countries.

Laser target designators may also be used on helicopters, warships, armored vehicles, and remotely piloted vehicles (RPV) and in various other applications. Their development has been rapid, and the number of designators, both those integrated in weapons systems such as in aircraft and those carried by soldiers, is gradually increasing.

One of the arguments sometimes put forward against the battlefield laser weapon concept is the difficulty of hitting the target with the laser beam. However, the results of the latest combat tests to date studying the efficiency of laser target designators prove that this is not the case. Placing an invisible laser beam on a small target is completely possible even from a fast-moving jet aircraft, and it is certainly much easier for a soldier on the ground, in a stationary position, to direct the beam at even smaller targets. The only important limitation is that the target has to be in the line of sight of both the target designator and the airborne munition seeking the target simultaneously. This, however, gives the enemy two visible targets at which to aim in order to knock out the weapons system. The target designator is especially vulnerable, because it will move relatively slowly, if at all, and must remain on longer. This limitation on the use of the target designators will not be a limitation in the case of laser weapons which depend only upon the effect of the beam itself and are on for much shorter durations.

BEAM RIDERS

Another way to use the laser beam for weapons guidance is to literally let the missile stay within the beam all the way to the target. Such a beam rider system differs from a target designator in that the missile can be steered during flight, and also a much less sophisticated detector is needed. The operator aims the laser beam at the target and then launches the missile, which has a rear-facing infrared sensor to determine the deviation between the beam and the missile. A great advantage of beam riding systems is that the beam is difficult to deflect by any electronic counter-measures, and, as the beam detector is at the rear of the missile, it is less vulnerable to overload by enemy activity.

The first operational beam rider system was the Swedish anti-air missile Rb 70, used in the daytime. This was soon followed by the day/night version, known as Rb 90. Another beam rider is the Swiss ground-to-air and antitank missile system, ADATS, and there are more systems in the R&D phase in other countries, including the high-velocity missile (HVM) system of the United States.

The first generation of beam riders used a semiconductor laser such as a pulsed GaAs laser, which operates in the near-infrared part of the spectrum at approximately 900 nanometers. New beam riding systems such as the ADATS use a CO_2 laser. There are several advantages associated with the longer wavelength of the CO_2 laser (10,600 nanometers). The transmission through the atmosphere is better, turbulence is not as much of a problem, and a higher average power is available, all of which means better performance in most situations and especially in bad weather. Another advantage of the CO_2 laser is its compatibility with FLIR systems in the 8,000- to 13,000-nanometer band. The CO_2 laser is presently the first choice for the laser guidance of the new American HVM system, which is designed as an antitank system.

The beam riding technique is on the battlefield to stay. There will undoubtedly be new beam riding systems fielded in the future, most of them using the longer wavelengths and more powerful outputs.

SIMULATORS

One of the difficult budget problems common to all armed forces is the cost of training with live ammunition. The cost of an antitank missile, for instance, may be in the neighborhood of $10,000, which means that a missile operator only can fire a very limited number of missiles during his training and peacetime duty—in some armies, perhaps none. Live firing also involves a risk of accidents, and the firing can only be done at an artillery range or in one-sided engagements against artificial targets. The restricted number of ranges suitable for firing missiles limits the possibility of using the right tactics and strategy, mostly because of a lack of space and safety considerations. For many years, this problem has been solved for rifles and machine guns by using blank ammunition during field exercise. However, the use of blanks does not give a very realistic simulation of the live munitions situation, and the simulation of more sophisticated ammunition, such as missiles and smart ammunition in this way is not possible. Moreover, the use of blanks does not give any indication of the gunner's ability to hit the target and often may not affect the response on future action by the targets after they are supposedly hit.

The cure for this age-old problem is to use laser systems which simulate the firing of missiles, projectiles, or rockets, evaluate the effects, and illustrate the results to both the gunner and the target. A laser is an ideal training aid for all sorts of firing, as a laser pulse will be sent to the target instead of a bullet, projectile, or missile. Basically, laser weapon simulators may be used for two training purposes—first, to give the soldier basic training with the weapon in question and, second, for training in two-sided field exercises where combat units with different weapon systems use their weapons against live targets. The ultimate training is when two large combat units, brigades or divisions, can fight each other in the field using laser weapon simulators on all levels from the infantry to tanks, helicopters, and aircraft. This is already being done in some countries today, including the United States.

In these laser systems, it is possible to simulate every single part of a live firing situation and, thus, ensure that the personnel trained in this way are doing everything exactly right. The cost of each laser shot is negligible. After the investment in the laser weapon simulator has been made, each soldier may literally be able to shoot hundreds and maybe even thousands of times, each time receiving complete information on what was done right or wrong. To a large extent, this is the state of the art in present training practice.

The laser used in the simulators is normally a very low power semiconductor laser. These lasers are small, easy to maintain and use, and relatively cheap. A semiconductor laser makes it possible to determine the three spatial coordinates of the simulated projectile and its position in relation to the target with very high precision. It is, of course, an absolute necessity that the laser used be completely safe to the eye. This is a requirement for two-sided exercises as well as for basic training, as most simulators work by reflecting the laser beam from the target using a retroreflector.

There are many laser simulation systems in use all over the world. A good example is the U.S. MILES system for small-arms training. It is used as a firing simulator on the automatic weapons of the infantry soldier. A target detector is fastened on the helmet as an indicator to tell the soldier when he is hit. Also, his weapon may be disabled when he is hit in a two-sided exercise. The MILES system is used for basic training, for training in squad and platoon level battle techniques, and as an integrated system in large-scale combat exercises. Tactical firing simulators for tanks are used in the basic training of the tank gunner and commander in using the tank gun and in tank-versus-tank battle training. The Swedish BT 41 is an example of a tank-mounted laser firing simulator. There are even simulation laser systems for other direct-line-of-fire weapons such as antitank missiles, anti-aircraft missiles, and guns. The development in this field is rapid, and the budget problems in many armed forces will certainly result in a demand for more systems that can reduce training costs and still maintain a high degree of proficiency in the various combat units.

Even in these laser simulator systems, as in the target designator systems, the soldier is required to aim a laser beam onto the target. This is done by using the standard fire control equipment (rifle sight, etc.) and other standard procedures. It is not very difficult to hit the target with a laser beam at normal battlefield engagement distances. If this simulator were a battlefield laser weapon, it would hit the target just as well.

LASER RADAR

The difference between laser radar and other military laser applications is somewhat blurred. Laser radar is the widely accepted name for a laser system that can give more information than range finding alone—for example, target velocity by Doppler effect analyses, scanning, and tracking. The laser tracker described earlier may fall within this definition and could be called a radar system.

In essence, a laser radar system uses a laser scanner to cover the required field of view. The scanner may be a rather simple mechanical device with rotating mirrors. The beam is scanned through the field of view in a pattern similar to that used on a television screen. The various objects in the field of view reflect part of the laser beam back through the optical system of the laser radar, where it is then analyzed. It could be said that the laser radar works like a laser range finder, measuring the range to every point in the field of view, thereby building up a complete picture.

One of the main advantages of laser radar over ordinary radar is its short wavelength, resulting in much higher resolution. This means that the laser radar is capable of not only detecting extremely small targets but also describing the subtle features of their motion. Thus, from the Doppler frequency changes in the reflected beam it is possible to discover and record extremely small movements such as the vibrations from a vehicle engine, the motion of the rotor blades on a hovering helicopter, or the combustion in an aircraft jet. Since many vehicles have characteristic vibration patterns or signatures, it may then be possible to identify the specific type of tank which is the target. Another important advantage is the possibility of building very small and compact laser radar devices that can easily fit into a fire control system rather than using a bulky conventional radar system which must be operated as a separate unit. The laser radar operates independently of the ambient light conditions and is unaffected by the absence of daylight. The laser radar is difficult to discover and counter. This, combined with the atmospheric attenuation of the beam, makes the laser difficult to detect and nearly impossible to jam.

Certainly, the laser radar concept has disadvantages. Perhaps, the most important one is the range, which is severely limited by difficult atmospheric conditions. If the weather is good and the air is clean, it may be possible with a 10–50-watt laser to reach an effective range of about 6 miles against a helicopter. On the other hand, with battlefield haze added to more typical weather conditions, the range may very well be no more than 3–5 kilometers, which is comparable to that of an FLIR system.

The laser of choice for radar seems to be the CO_2 laser. It is relatively efficient on fuel and has a rather good atmospheric transmission in smoke and fog. Also, it is suitable for coherence detection, which is a prerequisite for high sensitivity at the longer ranges. The wavelength for the CO_2 laser may be chosen in the band between 9,000 and 12,000 nanometers and is in the same band as the FLIR thermal sensors. This spectral overlap means that the two systems may be integrated in a single instrument with shared optics. FLIR and the CO_2 laser radar have about the same range under reduced visibility conditions.

The use of a coherent CO_2 laser in laser radar is very advantageous, as it is possible to determine a moving target's speed and vibration characteristics with high resolution while tracking and scanning. However, the most interesting attribute of the coherent CO_2 laser compared to the other alternatives is its longer range. If the maximum range under bad weather conditions is compared to that of another laser alternative—for example, the Nd:YAG laser with a wavelength of 1,540 nanometers—the result is very much in favor of the coherent CO_2 laser. A possibility exists that the use of a frequency-doubled coherent CO_2 laser at wavelengths between 4,500 and 5,500 nanometers could increase the range even more.

Another possible laser radar application has been demonstrated by Sandia National Laboratories in New Mexico. The system uses a gallium arsenide semiconductor laser diode of about 120 mW to produce a near-infrared beam which is scanned across and up and down to cover a field of about 30°. Range information is derived from the returned signals, and an image of any object in the field of view is displayed on a screen and updated four times per second. The range of this experimental system is only about 55 yards, which is quite short. More powerful lasers that can provide almost 5 watts are available and could be used for experiments on larger ranges. Future applications could include automatic target recognition or more precise aiming for conventional weapons. Use in collision avoidance systems and drone or robot vehicle control is also a possibility.

There are many possibilities for laser radar applications. Terrain following and navigation aids for aircraft, helicopters, and missiles can be made much more efficient with laser radar than they are today with ordinary radar. This concept may allow covert operations as part of the Cruise Missile Advanced Guidance Program (CMAGP) in the United States. One company is now developing a laser radar which is said to provide improved navigation and targeting capability for present and future cruise missiles.

If the laser radar is combined with an FLIR system, this may mean a breakthrough for air-to-ground automatic targeting systems. The FLIR alone may give only the outline of the target against a sometimes complex background; adding laser radar provides range, reduces background, and separates moving targets. The pooling of information from the FLIR and the laser radar may more easily give a correct identification and separation of multiple targets at the same time. It is also possible to use a laser radar system as a target designator or as a source for a beam rider. This is important for laser weapons aimed at sensors, optics, and eyes, as they can then easily locate and pinpoint the strong reflections from optics at long ranges.

A new German concept is a sensor-fused, fly-over dispenser designed to attack armored vehicles from a low altitude while flying over the target. The sensor combines laser radar to determine the surface contour together with a radiometer to detect any identifying reflective radiation and an infrared sensor to determine the temperature profile.

Laser radar is still in a research phase, and it will take some time before operational units are fielded and used in any numbers within any armed forces. However, there are certainly many applications possible in antitank, anti-aircraft, air-to-ground, and sea warfare. The small size and versatility of laser radar equipment could make it standard equipment in the 1990s. The combination of a laser radar system with a laser weapon offers the possibility of detecting optics, and the sensor or eye behind the optics, and automatically firing the weapons against the sensor or eye behind the optics.

There are two other military applications that should be mentioned here—Light Detection and Ranging (LIDAR) and laser use for bathymetry or depth measurements. LIDAR is a very important tool for investigating and analyzing different characteristics of the atmosphere. The beam from a pulsed laser is directed into the atmosphere, where it is scattered by different types of molecules. Some of the radiation is reflected back to the LIDAR, where the results can be analyzed. Military units may use LIDARs to detect chemicals, measure the wind parameters for artillery, or determine the local weather. Some atmospheric-borne chemicals may be detected at long distances by a coherent CO_2 laser.

It is also possible to use a laser for bathymetry to measure and to map the characteristics of the bottom of an ocean, a lake, or a river if the water is not too deep. By using a pulsed green laser, for example, a copper vapor laser, it is now possible to reach a depth of 30–40 meters if the water conditions are favorable. A short pulse is emitted which first hits the surface of the water, where a part of the beam is reflected and can be detected. The remaining part of the beam is transmitted through the water and back to the receiver. It is, thereby, possible to measure the depth and map the bottom by scanning the laser beam. This technique may be possible for military units to use when trying to detect marine mines and even submarines.

OTHER MILITARY APPLICATIONS

Range finding, target designation, and radar analogs are obvious military applications. There are many other possible military applications for lasers, some of which also can be part of a weapons system.

The laser beam may be used for long-range communication purposes with the information carried by the modulation in intensity of the laser beam. One example of this is the U.S. Navy Satellite-Based Submarine Communications Laser Program (SLCSAT). This is based on a Raman-shifted xenon chloride gas laser emitting a multijoule pulsed beam in the blue part of the spectrum. The wavelength has been reported to be 459 to 449 nanometers for submerged reception. Current efforts are directed toward overcoming the problem of the size and bulk of the laser and making a 500-million-shot life possible.

On the battlefield, electromagnetic jamming of present-day radio and wire systems is an increasing threat. Laser communication through air or optical fibers offers what may be the only really secure solution. The Yugoslavian laser communication device RLK2 is representative of many hand-held, compact, and jam-proof devices. It weighs about six pounds, emits at 905 nanometers, and can transmit voice and data (10,000 bits/second) for a maximum range of 4.4 miles through the atmosphere. It may operate for eight hours on one package of batteries. Such devices will be used in many armed forces in the future.

The most important laser for present-day communication purposes is the semiconductor (diode) laser, which is very compact and efficient. Optical fibers are used instead of the atmosphere for transmission of the laser beam. This is much more efficient, as billions of bits per second can be transmitted through the optical fiber. Extensive military use of optical fibers for many different purposes will almost certainly force the development of even more efficient and less expensive semiconductor lasers.

Ring laser gyros are crucial for the performance of many navigation and guidance systems for high-technology missiles, aircraft, and helicopters. They provide direction, pitch, and yaw reference data. A ring laser gyro is, compared to a conventional gyro, small and light and has no moving mechanical parts. This means that the gyro is very suitable for rough environments and can easily cope with rigid military requirements. The laser used is a standard HeNe laser, and the ring is not a circle but rather a triangle or square with mirrors at each corner to reflect the beam, which is split in two and sent in opposite directions around the circumference. Velocity differences are then measured between the light beams passing in opposite directions around the "ring" so that any rotation of the instrument in the plane of the ring can be detected to a very high degree of accuracy.

A major development in laser technology for support of the infantry is the laser sight or laser pointer. This is somewhat analogous to the target designator. Small arms, rifles, etc. can be aimed at the target by using a laser to produce an intense visible spot on the target. The system is an add-on piece of equipment, and it can line up with the barrel in such a way that the laser spot is in the same place as the predicted point of impact of the bullet at the desired range. It is possible to change the alignment in accordance with the type of ammunition and variations in range. Laser pointers were initially developed to provide an extremely fast and accurate point-and-shoot capability in daylight for police and security people without endangering innocent bystanders. These systems mostly use a ruby laser producing a highly visible red spot on the target. For military purposes, it would be advantageous if a laser pointer could have an invisible spot seen only by the gunner and which could be used even at night. This has been made possible by using the combination of passive night vision goggles and a laser pointer based on a laser in the near-infrared part of the spectrum, which is thus invisible. Development has been very rapid, and a number of such devices are already on the market. A recent review in a military magazine describes more than 10 laser pointers from different countries. One example is the LM-18 from the German company Euroatlas, which is designed to be used together with night vision goggles. It weighs about three-quarters of a pound and operates in the 800–870-nanometer range with a spot size of 2 inches at 55 yards range. The maximum ranges claimed for different pointers vary from 240 to 360 yards.

This laser sight technology is another example of a military laser system where a soldier has to point and hold the laser at the adversary, a task which is certainly possible not only with a light machine gun combined with a laser sight but also with an antipersonnel laser weapon.

There are many more military laser applications involving techniques such as measuring air and liquid pressure using a semiconductor or a HeNe laser. Another use for military lasers is to find the fuse to set a bomb or missile off automatically at a predetermined moment.

SUMMARY

The military use of laser technology is well established, and the laser is now a standard tool for many soldiers in many different weapons systems. The use of lasers has to some extent already influenced military tactics and doctrine. The laser has already solved many problems in ranging, tracking, and target designating and is an integrated part of many high-technology fire control systems. The use of laser simulators has made both basic and advanced training more efficient and in many cases even cheaper. It is obvious that the number of military laser applications will increase steadily in the future.

Against this background, the question of the morality of laser weapons is very interesting. The armed forces in most countries are now getting used to pointing lasers at the enemy for ranging, tracking, and designating. If laser weapons turn out to be a realistic alternative or complement to conventional weapons, their use against soldiers themselves will be a very small step from how they are already being used on the battlefield today.

THREE

Laser Safety

The laser has become a common tool of civilians and soldiers all over the world. Many lasers, perhaps most of them, are in some way dangerous to people. For several reasons, it is our eyesight that is most threatened, but there are many other dangers to deal with as well. Laser safety is a very complex problem. This chapter will provide some basic facts about the problem and will use military lasers as examples, keeping the laser weapon question particularly in mind.

LASER HAZARDS

The use of lasers almost always carries with it some kind of danger, either at the laser site itself or wherever there is a direct, reflected, or scattered laser beam. At the laser site, it is not only the actual laser beam which can be dangerous, but electrical, chemical, and other hazards exist as well. Most laser power supplies can cause severe electrical shocks, possibly even electro-cution. Furthermore, many highly explosive and toxic substances are used in solid, liquid, or gas form to power laser cooling systems. The danger of being exposed to these substances is most prominent in laser laboratories and factories. In addition, there may be serious toxic effects from the vapors released into the air during the processing of laser materials. Electrical, chemical, noise, and other related hazards are serious matters to consider, but they are not associated with laser weapons and, therefore, will not be explored further.

In and around most laser operations, there is always the possibility of a fire hazard. Flammable material, such as paper, may be set on fire by a CW laser operating with an output power above 0.5 watt. The effect is, of course, similar to that of a high-energy laser weapon designed to burn holes in and set fire to different targets.

For many reasons, it is useful to divide the hazards from laser beams into two main groups: those to the eye and those to the skin. This chapter will describe in some detail why the eye is by far the organ of the human body that is most sensitive to laser radiation. The eyes may be severely damaged and even permanently blinded by rather low energy laser beams, while the skin is not nearly as sensitive. To get severe skin burns in the visible and infrared part of the spectrum, it is normally necessary to use a very high energy laser beam which delivers, at least, several watts per square centimeter (W/cm^2) to the target.

Safety threshold limits both for the skin and eyes are well defined and have resulted in very strict safety regulations. Many different national and international regulations exist, but they are all based largely on the National U.S. Standards and particularly the American National Standards Institute's Z-136 Laser Safety Standards and its revisions.

LAZER HAZARDS TO THE EYE

To understand laser hazards to the eye fully, with their implications for laser safety requirements and the possibilities of anti-eye laser weapons, discussed in later chapters, it is necessary to begin with a very short overview of the anatomy of the eye.

The eye is a very complex and precise optical system. It is approximately 1 inch (25.4 millimeters) in diameter. The general structure of a standard left eye is indicated in Figure 3.1. When light enters the eye, it first passes through the cornea, a living tissue exposed directly to environmental elements. The cornea is protected from drying out by the tear film. It is responsible for the major part of the focusing of light rays on the rear of the eye. Once the light passes through the cornea, it then enters the aqueous, which, as its name implies, is essentially composed of water. The pupil, the next part of the eye through which the light passes, has its size adjusted by the iris—the colored part of the eye. The size of the pupil is adjusted from approximately $\frac{1}{12}$ to $\frac{1}{4}$ inch (2–7 millimeters) in response to the average brightness of the light. This is an important factor to consider when the possibility of

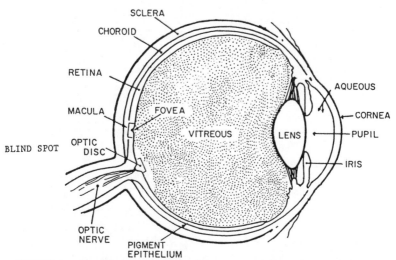

FIGURE 3.1. Simplified cross section of the eye, identifying the principal structures. This cross section is a horizontal slice of the left eye viewed from above. (Adapted from D. Sliney and M. Wolbarsht, *Safety with Lasers and Other Optical Sources*, Plenum Press, New York, 1980.)

damage to the eye is discussed. The wider the pupil is, the greater the amount of light that passes into the eye.

The lens, directly behind the iris, is colorless, as it does not absorb throughout the visible region of the spectrum. After the light has passed the lens, it goes through the vitreous body, which is a colorless gel. The vitreous body is attached to the ciliary body and to the retina at several points. After passing through the vitreous body, the light finally hits the retina. Changes in the focus of the light rays entering the eye are controlled by the ciliary muscles, which change the shape of the lens. Accordingly, this allows far or near objects to be focused properly on the retina.

The retina consists of several very complex layers of nerve cells and is, in fact, an extension of the brain. The outer surface of the retina is covered by a single layer of cells called the retinal pigment epithelium (RPE). Just inside this layer are the photo-receptor cells, consisting of two types, rods and cones. Every eye has approximately 125 million rods and 7 million cones. The rods are large, cylindrical cells that are very sensitive to low intensity levels of light but are not used in color vision. The cone cells are smaller, occurring mostly in the central portion of the retina. They are responsive to higher light levels and are used in distinguishing colors. The adjacent layers in the choroid, outside the retina, are also of special interest in an evaluation of retinal injury from light sources.

When a person looks directly at an object, it is seen in fine detail. At the same time, the surroundings are also seen, but with less clarity. To see the surroundings in detail, the whole scene must be scanned. This is because when the eye is aimed most directly at an object, only a small area in the center of the retina, called the fovea, is being used. The fovea is very densely packed with cones, possibly in order to discriminate the fine details of the image. In any case, the fovea is responsible for the highest acuity. The fovea is the center of a larger area called the macula, or yellow spot, where the vision is still very good. In this area, there may be more than 4 million cones. It is important to remember that the foveal area is responsible for the majority of high-acuity viewing, and, indeed, when the fovea or the central macula area does not function, a person is very severely visually disabled. This can be appreciated at night, when the fovea and most of the macula lose their ability to detect details, as the cones which are responsive only to high levels of light do not work. Darkness eliminates the fovea completely, as it has no rods, along with the central macular vision. Therefore, the outer portion of the macula is responsible for the best vision under these reduced illumination situations. The distribution of visual acuity for both rods and cones is shown in Fig. 3.2. It should be kept in mind that at least 20/40 vision is necessary for driving an automobile or rapid reading.

The eye does not transmit the whole range of electromagnetic radiation. This is a basic fact that has a bearing on both laser safety and laser weapons. Figure 3.3 shows that infrared radiation (beyond 1,400 nanometers) as well as far-ultraviolet rays (wavelengths

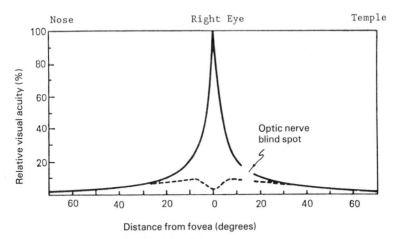

FIGURE 3.2. Variability of visual acuity as a function of visual angle or distance away from the central portion of the eye, the fovea. The solid line is for a bright-light (photopic) response, and the dashed line for a dim-light (scotopic) response. (Adapted from D. Sliney and M. Wolbarsht, *Safety with Lasers and Other Optical Sources*, Plenum Press, New York, 1980.)

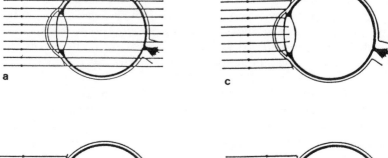

FIGURE 3.3. Schematic diagram of the absorption of electromagnetic radiation in the eye. (a) Microwaves, X rays, and gamma rays all pass through the eye with little change. (b) Far-ultraviolet and far-infrared radiation are absorbed in the surface layer of the cornea, whereas (c) near-ultraviolet radiation is absorbed mostly by the lens. Only visible and near-infrared radiation (d) are focused by the optical elements of the eye on the retina. (Adapted from D. Sliney and M. Wolbarsht, *Safety with Lasers and Other Optical Sources*, Plenum Press, New York, 1980.)

less than 315 nanometers) do not get past the cornea. Most near-ultraviolet radiation (315–400 nanometers) is absorbed harmlessly in the lens. This means that neither the laser energy from a CO_2 laser at 10.6 micrometers nor that from a nitrogen laser at 305 nanometers can reach the retina. On the other hand, really high energy radiation such as X rays and gamma rays can penetrate through the whole eye but does so without much damage. In contrast, visible (400–700 nanometers) and near-infrared (700–1,400 nanometers) laser energy can be focused to a point on the retina and will cause damage if intense enough. For this reason, visible and near-infrared energy are together labeled as the retinal hazard region.

Several details about the retina are important to understand. Before light reaches the rod and cone layers in the retina, it has to pass through the inner retinal layers of blood vessels, nerve fibers, and nerve cells. However, in the fovea, all of these inner layers are missing, and, in this area, the light has direct access to the cones and rods, in keeping with the high resolution possible in this area.

A very important characteristic of the retina is its ability to adapt to light and darkness. A bright light immediately lowers the eye's sensitivity to light—only lights of similar brightness can be seen after exposure to a bright light. This effect is often referred to as flash blindness. That is, after exposure to a bright light, at first it is not possible to see anything which is less well illuminated, but, after some time, even a low light level will allow various objects to be perceived. The whole process may take a long time, anywhere from 20 to 30 minutes. The time course of the retina's adaptation to the dark is shown in Fig. 3.4. At night, colors are not seen because there is only enough light to stimulate the rods even though the pupil opens up to admit more light into the eye. The time required for dark adaptation is of great importance when it comes to any laser weapons bright enough to cause flash blindness.

The visual field from each eye is roughly 50 degrees upward, 80 degrees downward, 60 degrees nasally, and more than 90 degrees temporally. This means that the combined visual field from both eyes is more than 180 degrees. There are two blind spots in this field, one for each eye, which correspond to the area in each retina covered by the entrance of the optic nerve. The two blind spots are far apart in visual space, and, since the brain has a tendency to fill in blanked-out areas in one eye by the contribution from the other eye, these blind spots are not normally noticed.

When discussing laser safety and the eye, it is necessary to differentiate between the effects inside the eye within the retinal hazard region (400–1,400 nanometers) and the effects on the outside of the eye from those laser beams that do not reach the retina.

The Retinal Hazard Region

The ability of the eye to transmit laser radiation to the retina varies significantly within the retinal hazard region. The transmission is at its highest in the visible part of the spectrum, drops in the near-infrared region near 1,000 nanometers, has another peak around 1,100 nanometers, and then drops sharply. The transmission of the eye is, thus, very dependent on the wavelength, as shown in Fig. 3.5. For example, green is almost twice as well transmitted as some parts of the near infrared. However, the difference in transmission between a frequency-doubled (green) Nd:YAG laser at 532 nanometers and an ordinary (infrared) Nd:YAG laser at 1,064 nanometers is only about 30%.

Because of the unique collimation and coherence characteristics of the laser beam, it is seen by the eye as if it were coming from a point source. Thus, the collimated laser beam is collected by the

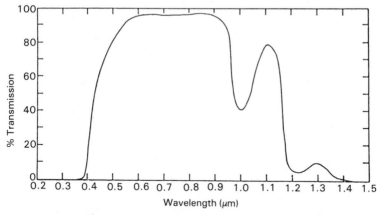

FIGURE 3.5. Spectral transmission of the ocular media. This chart shows the relative ability of visible and near-infrared radiation to reach the retina. Ultraviolet (less than 0.4 micrometers) and mid- to far-infrared (greater than 1.4 micrometers) radiation do not penetrate the eye sufficiently to reach the retina. (Adapted from D. Sliney and M. Wolbarsht, *Safety with Lasers and Other Optical Sources*, Plenum Press, New York, 1980.)

optics of the eye, transmitted through the ocular media, and focused to a very small point, 20 to 50 micrometers in diameter, on the retina. The effect of this concentration is to magnify the brightness of the light by a factor of approximately 100,000. If the diameter of the laser beam is larger than that of the pupil, external magnifying optics such as a telescope will make the effect even worse. Thus, a pair of binoculars, 7 × 50, with a light transmission of 50% will increase the energy focused on the retina by the ratio of the area of the objective lens (50 millimeters in diameter) to the area of the pupil (7 millimeters in diameter) divided by the brightness factor, which in this case amounts to 50/2 or a 25-fold increase in the energy reaching the retina. Thus, an individual will increase the risk of damage to the eye by adding 25 times more energy to the retina when looking through binoculars than when using the naked eye.

This additional hazard from an optical aid is, of course, of the greatest importance when it comes to the fielding and use of anti-eye laser weapons. A soldier using magnifying optics may not only be much easier to blind but may also be a more valuable target than a soldier with a naked eye. Tank gunners, artillery fire controllers, missile operators, commanders, and others all use magnifying optics in critical moments on the battlefield, and their optical systems may then be detected and identified by the characteristic reflections and exposed to hostile laser radiation by the enemy.

The hazardous effects of a laser beam that is transmitted through the eye are, in the vast majority of cases, limited to the retina. The effect upon the retina may range in severity from a temporary reaction without residual pathological changes to permanent blindness. The smallest observable reaction may be a whitening of the retina. However, as the retinal irradiance is increased, lesions occur which progress in severity from swelling (edema) to burning (coagulation) and then bleeding (hemorrhage) as well as additional tissue reaction around the lesion. Very high retinal irradiance will cause gas bubbles to form near the site of absorption. These gas bubbles can disrupt the retina and in some cases alter the physical structure of the eye.

The retina itself is not much more sensitive to laser damage than any other parts of the body. The level of energy that may cause severe damage to any part of the body is between 50 and 500 millijoules per square centimeter for a short pulse. It is only the optical concentration of the energy by the optics of the eye that makes a low-powered laser capable of damaging the retina specifically rather than the rest of the eye or body.

The most important part of the eye for vision is the macular area and, in particular, the fovea centralis, which is densely packed with cones. If the laser beam causes a retinal burn of any size in this area, the result is permanent loss of fine-detail vision sufficient to cause legal blindness, and no treatment is possible. Of course, much vision is still present but not enough to read rapidly, drive an automobile, or do any visually demanding task.

If the retinal burn does not affect the macula, the visual impairment may not be very serious. The focused laser beam can leave a small wound on the retina, which may be seen by the person involved as a small dark spot in his visual field. In many cases, such a spot will not be noticed at all. The laser exposure itself is more or less painless, and the dark spot very small. However, if the energy level is sufficient, such laser exposure can cause a vitreal, infraretinal, or subretinal hemorrhage inside the eye. Vision will be obscured by the blood blocking the optical path between the lens and the macula or by the blood elevating the retina and detaching it from the pigment epithelium. Such a hemorrhage is a serious injury requiring immediate medical attention. Any damage to vision may be permanent, although some treatment is possible. This treatment is complicated and involves sucking blood from under the retina or removing the vitreous body. Both operations require exceptionally clean operating rooms and are extremely difficult, if not impossible, to attempt on a battlefield or in other field-type hospitals. At present, it also appears that if the treatment is not completed within a certain time after the injury, perhaps a week or less, the damage will most certainly be permanent and may even get worse.

The laser's capability to cause various types of eye damage at many different distances has been very well documented in numerous animal experiments. This rather large body of data has also been compared to data on the real injuries of humans in a number of actual laser accidents.

Some experiments performed in the United States on rhesus monkeys who were exposed to laser radiation in order to determine the threshold levels for serious eye damage can be summarized as follows. A pulsed green laser could cause a retinal burn at about two miles and a retinal hemorrhage at a quarter of a mile. The use of binoculars increases these distances by a factor of about 4. The corresponding distances for a deep red (ruby) laser of the same energy were slightly shorter than for a green laser, while the infrared laser was much less effective than both. The details are given in Table 3.1. In a Swedish study, the possibility of a standard ruby range finder causing a vitreal hemorrhage in the eye at battlefield distances was investigated by using the eyes of anesthetized pigs to simulate observation with the naked eye or through magnifying optics. The report on the data concludes that a standard military ruby range finder under normal atmospheric condi-

TABLE 3.1.   Laser Injury Experiments on Rhesus Monkey Eyes[a]

| | | Distance (kilometers) at which damage occurs | | | |
| | | Vitreal hemorrhage | | Retinal burn only | |
Color	Wavelengths (nanometers)	Without optics	Optics	Without optics	Optics
Green	530	0.65	2.9	3.2	10.5
Deep red	690	0.47	2.4	1.7	6.6
Infrared	1,060		0.25	0.61	3.0

[a]In all cases, a Q-switched laser was used with 100 millijoules in each pulse. The beam had a divergence of 0.25 mradians. The optics used were 7 × 50 M17 binoculars. A 7-millimeter pupil was used for all exposures, both with and without optics, and a clear, nonturbulent atmosphere was present during the exposures.

tions can cause vitreal hemorrhages in soldiers employing binoculars on the battlefield at engagement distances up to, at least, half a mile. Similar data for humans is given in Fig. 3.6.

Hundreds of injuries have resulted from accidental exposures to laser radiation. Laboratory or experimental laser setups accounted for many of these, but an ever increasing number are due to exposures during military training and operations. Most of the accidents were caused by invisible radiation in the near-infrared region and usually involved the central retina, the macula, and occasionally even the fovea. Outdoor accidents produced similar injuries in both eyes, but most accidents occurred indoors, or at close range, and involved one eye only. Only limited, if any, recovery was reported, and the accidents were normally quite traumatic psychologically to the patient.

The mechanism of tissue damage discussed so far has been limited to the thermal effects of the laser beam in cases where the laser energy is transferred to the target tissues at a faster rate than that at which the heat can be dissipated safely. However, the thermal effect is not the only hazard associated with lasers. There are two other basic mechanisms of damage: ionization and photochemical changes (see Fig. 3.7). Ionization may be described as pulling the molecules apart and occurs as the pulse duration is decreased and the photon density increased. It is similar to spark generation. Photochemical damage can occur because the chemical processes in the body tissues can be influenced by ultraviolet radiation and blue light. Sunburn is a good example. Damage to the cornea and retina can result after exposure to long pulses or CW beams.

Some of these effects are not too well understood at present, and much work still remains to be done before the damage mechanisms of the laser beam are completely predictable. However, there seems to be no doubt that a low-energy anti-eye laser weapon can disable a person visually even if it is dependent only upon thermal effects within the retinal hazard region.

What has been said so far about the action of lasers on the retina has been focused on the injury aspects. However, another

ND:YAG (1064 nm) range finder

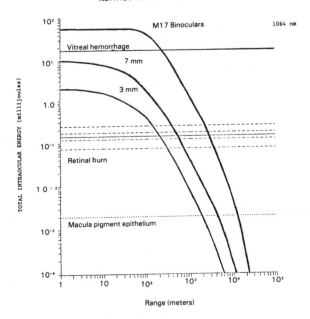

ruby (694.3 nm) range finder

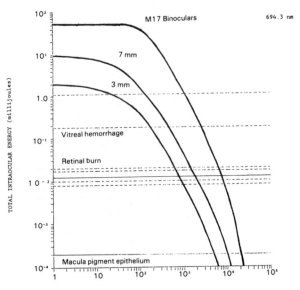

effect from a laser has to be taken into consideration when the possibilities of laser weapons are discussed. Before the energy levels are sufficient to cause a retinal burn, the laser beam may very well cause uncomfortable glare or flash blindness. These effects are very dependent upon the circumstances—whether the eye is dark adapted or not, the wavelength and pulse length of the laser beam, the ambient light levels, and various other environmental influences. The prospect of using a laser beam for flash blinding of pilots, tank crews, and other key soldiers on the battlefield must be considered as a possible laser weapon application for use at night. This possibility will be discussed in greater detail in Chapter 6.

Laser Effects outside the Retinal Hazard Region

The retinal hazard region covers the spectrum from 400 to 1,400 nanometers and includes the visible and near-infrared parts of the spectrum. As already mentioned, shorter wavelengths in the near ultraviolet are absorbed mainly in the lens, and the even shorter far-ultraviolet wavelengths are absorbed mainly in the cornea. Longer wavelengths in the mid-infrared region are also absorbed in the cornea. Thus, there can be harmful effects to the eye from laser exposures even in the parts of the optical spectrum outside the retinal hazard region.

The excessive absorption of intermediate ultraviolet radiation by the cornea causes ultraviolet photokeratitis. This is a very painful but temporary injury, often called snow blindness or

---

FIGURE 3.6. Laser ocular effects at various ranges from military range finders under tactical conditions. The various types of eye damage—retinal burns, vitreous hemorrhages, etc.—are shown as a function of the energy and the distance from the range finder. The curves shown are for a 3-millimeter pupil, a 7-millimeter pupil, and a 7-millimeter pupil with optically aided viewing using M17 binoculars. The two graphs are for a Nd:YAG (1064 nm) and a ruby (694.3 nm) range finder, respectively. (Adapted from B. Stuck, Symposium on Medical Surveillance, September 8–9, 1982, Aberdeen Proving Ground, Maryland, p. 30.)

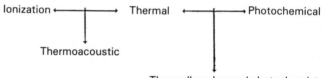

FIGURE 3.7. Types of interaction of laser energy with the eye and other biological tissues. Only the thermal and thermoacoustic modes of interaction are important with present-day antipersonnel laser weapons.

welder's flash. The eyes are very sore, and the movement of the lid over the cornea in blinking is very painful. However, although this effect lasts only one or two days, the person is effectively blinded during this period. For comfort, the eyes must remain shut and should even be bandaged. Extremely high exposure levels may permanently damage the cornea and possibly the lens behind it. The photochemical effects of ultraviolet radiation are not yet fully understood. However, as most of these effects require that the eye be subjected to high levels of laser radiation for a comparatively long time, half a minute or more, it seems unlikely that these effects in the ultraviolet part of the spectrum could form the basis for a laser weapon.

In the mid- and far-infrared region, the possibility for absorption in the cornea, especially for wavelengths longer than 2,000 nanometers, is very high. Therefore, the cornea is very susceptible to damaging heat during exposure to mid-infrared radiation. If the energy level of the beam is high enough to cause corneal heating, this will produce immediate and severe pain and automatically trigger the blink reflex. The cornea is quite sensitive, and an elevation of only 20°F will cause a pain response. The question is whether or not sufficient thermal energy would be absorbed in the cornea to cause injury in the short time before the blink reflex is activated. The lids are much less sensitive to damage because the circulating blood carries away the heat and a large amount of the laser beam is reflected.

The infrared lasers that may be used to injure the cornea are CO_2, hydrogen fluoride (HF), deuterium fluoride (DF), and CO. Such lasers with an output power of more than 10 watts per square centimeter could deliver at least 0.5 to 10 joules per square centimeter to the cornea before the blink reflex gives any protection, as shown in Fig. 3.8. Existing infrared lasers can certainly damage the cornea before any head movement can occur. Research has shown that thermal injury to the cornea produces a white spot or an opacification of the surface. The injury is extremely painful and needs immediate and well-qualified medical care. The severity of corneal burn injuries from laser exposure can be compared to that of burns and injuries resulting from the ignition or explosion of flammable objects.

LASER HAZARDS TO THE SKIN

The fact that the threat to the eye, especially in the retinal hazard region, is caused by extremely low energy lasers and, thus, constitutes a more frequent and obvious hazard has placed hazards to the skin as secondary to those to the eye. However, if the energy level is sufficient, the skin is a much larger target than the eye. It is also possible to burn the skin indirectly by setting fire to nearby objects with a high-energy laser. Obviously, widespread burn injuries are certainly more life-threatening than eye injuries.

Lasers can have several important effects on the skin. The thermal effect is the most significant one. Burn injuries are divided into three basic groups. A first-degree burn is a very superficial reddening of the skin, a second-degree burn produces blistering, and a third-degree burn, the most severe kind, destroys the entire outer layer of the skin. The irradiance necessary to cause a first degree burn is 12 watts per square centimeter; for second- and third-degree burns, the necessary irradiance is 24 and 34 watts per square centimeter, respectively. If the exposure

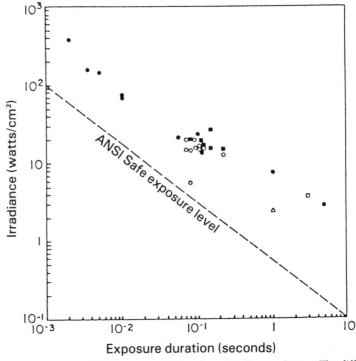

FIGURE 3.8. Threshold for corneal injury from CO_2 laser radiation. The differences between data points at the same exposure duration are due largely to the use of different corneal image sizes. The data points are from several laboratories and fit a thermal heat flow damage model quite well.

duration is shortened, the irradiance required to give a third-degree burn is significantly increased. Laser injury thresholds for the skin are dependent on the wavelength of the laser as well as on the pigmentation of the skin. Dark skins absorb more and thus get hotter for the same laser energy. For long exposures, the energy levels necessary to produce injury are highly dependent upon exposure duration. It is possible for high-energy lasers to produce significant burns within an exposure period of less than one second.

Most areas of the human skin are normally covered with clothing, and the total area of exposed skin may be rather small. A soldier on the battlefield, aware of the threat from laser exposures, will be rather well protected as long as his uniform or the immediate environment is not set on fire. However, it has to be recognized that, even in a protected state, burn injuries to the eyes will probably still be a problem. In the far-infrared and ultraviolet regions of the spectrum, where the laser energy does not reach the retina, corneal injury thresholds are approximately the same as for skin injury. Therefore, laser burns to both the exterior of the eye and skin are possible, but these do not seem to be important threats at the moment.

SAFETY REGULATIONS

The laser hazard from industrial applications and general public exposures has led to very detailed safety regulations for the use of various lasers. In the United States, the formulation of widespread standards started in the sixties at the same time that ruby range finders were introduced into the military services. In 1969, the American National Standards Institute (ANSI) initiated the first work on a comprehensive standard for the safe use of lasers. This resulted in a document published in October 1973, which has subsequently been revised several times. This work forms, more or less, the basis for laser safety standards all over the world.

The establishment of the threshold levels for different laser injuries is basic to the whole question of laser safety. The threshold level is an exposure value below which adverse changes have a low probability of occurrence and no significant risk exists. There is always some question about what the actual value of the threshold is, because it varies not only with the wavelength and exposure duration but also with the individual. The value of the threshold may be set by using a statistical analysis to determine a certain damage probability (usually 50%) and then setting the safety level at a selected level of probability below this, usually the 0.01 or 0.001% level. This energy/power level is often a factor of 10 below that for the 50% damage point. In order to calculate a correct threshold level, it is also necessary to try to simulate some kind of worst-case situation—when the eye is hit in its most sensitive part and takes in as much laser light from the laser in question as is possible under the circumstances.

Lasers are divided into four hazard classifications defined by the ANSI standard to assist in choosing the proper control measures for their safe use. The following is an outline of the basic classification plan.

Class 1 lasers are those not capable of emitting hazardous radiation under any operating or viewing conditions. Most lasers used in supermarket scanners and compact disc players are in this class.

Class 2 lasers are continuous-wave or pulsed types in the visible (400–700 nanometers) range which can be safely viewed long enough to identify and avoid the emitted light, much like one avoids directly looking at the noonday sun. However, precautions are required to prevent continuous staring into the direct laser beam. The familiar red HeNe lasers found in high school optics laboratories and used in lecture halls as pointers are in this class.

Class 3A lasers are normally not hazardous unless viewed with magnifying optics. They are used at all large construction sites by surveyors and to ensure that walls and drain pipes are placed properly and at the correct angle.

Class 3B lasers are potentially hazardous if their direct beams are viewed by the unprotected eye, but they do not cause hazardous diffuse reflections. Care is required to prevent direct beam viewing and to control specular or mirrorlike reflections.

Class 4 lasers are those lasers capable of producing hazardous diffuse reflections. They are also fire and skin hazards; their average output power is above 500 milliwatts.

It is obvious from the description of the various lasers that candidates for laser weapons will be found almost exclusively among class 3B and 4 lasers.

It may be of interest to consider what the safe distances are for current military laser devices. This will give an indication of the size of the hazardous area around each device. Each specific laser has a safety distance of its own based upon its output properties. The acronym used to describe this distance, in the U.S. regulations at least, is NOHD (nominal ocular hazard distance).

The NOHD is calculated to determine at what distance an unprotected person can stand directly in the beam and be exposed momentarily without being injured. The use of magnifying optics must be taken into account, because they will markedly increase the NOHD. A 6-mile NOHD may be increased to 50 miles if an individual looks into the laser with optics that magnify 13 times.

It should be remembered that the laser may be hazardous to the eye even beyond the NOHD if the laser is viewed or stared at for a prolonged time. The NOHD is calculated based on momentary viewing only. As can be concluded from Fig. 3.6, the distances necessary to reduce the laser energy to safe levels are really very long and may be difficult to find at any places other than military ordnance target ranges and training areas which can be securely closed to unauthorized persons, aircraft, and ships. Even within thoroughly supervised areas, it may be necessary to aim the lasers against natural or artificially constructed backstops to the laser radiation, because these areas are often too small to contain the NOHD for the particular laser. Laser backstops are opaque structures or terrain such as windowless buildings, hills, or a very dense line of trees.

It is also necessary to take into account the problem of accuracy in aiming the laser. There must be a buffer zone added to the hazardous area, and the size of this zone will be different for a stationary laser as compared to a hand-held laser. Also, the possibility of reflections from the target area is a problem that has to be very carefully handled. A specular or regular reflection is caused by mirrorlike surfaces such as windows, optical surfaces, greenhouses, still ponds, or road signs covered with reflective coating. All such reflections have to be avoided; therefore, such surfaces must be removed or covered, or some provision must be made for masking or shielding their reflection. A diffuse reflection which takes place when different parts of a beam incident on a surface are reflected over a wide range of angles is not that dangerous. However, for some lasers, there may be a hazardous diffuse reflection area (HDRA), which is typically less than 10 yards from the reflecting surface. An example of a laser danger zone is shown in Fig. 3.9.

The laser beam is decreased or attenuated by some atmospheric conditions, and this is a factor that should be considered when the NOHD is longer than a few kilometers. Atmospheric attenuation is mainly dependent on the sum of three different effects: large particle scattering, molecular scattering, and absorp-

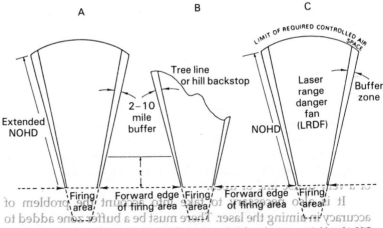

FIGURE 3.9. Laser range safety fans. Laser range safety fans are used by the U.S. Army to indicate the nominal ocular hazard distance (NOHD). The NOHD is normally terminated by a backstop. The unterminated NOHD depends on beam expansion and atmospheric attenuation. In case A, the NOHD is a line-of-sight fan parallel to the ground and would only be used when there is no backstop. Case B is the more usual situation where a backstop is established by a hill or tree line. In case C, the fan is perpendicular to the ground and is applied to airspace hazards.

tion by gas molecules. Large particle, or Mie, scattering is the dominant factor in the visible and the near-infrared part of the spectrum, where the particle size of the atmospheric contaminants is larger than the wavelength of the laser light. Molecular or Rayleigh scattering by oxygen, nitrogen, and other molecular constituents of the atmosphere is the most important factor in the ultraviolet, and, in these cases, the molecular size is much less than the wavelength. The contribution of absorption by gas molecules and other particles to attenuation is most important in the infrared region of the spectrum.

The molecular scattering of the laser beam increases at shorter wavelengths. However, this effect is not substantial over short distances. A normal and clean atmosphere is relatively transparent to the argon laser beam (blue), the ruby laser beam (red), and the Nd:YAG beam (near infrared). If the NOHD calculated for vacuum transmission is known and compared to the NOHD compensated for the ambient atmosphere (see Fig. 3.10), it may be concluded that, for low-energy lasers, the atmosphere even at battlefield distances up to 10 miles, at least, is not a big problem. A ruby laser beam might be attenuated as little as 10% at 6 miles. High-energy lasers are much more heavily dependent on the weather conditions, especially rain, snow, dust, and smoke. This will be discussed further in Chapter 5. However, what may be concluded from laser safety calculations is that most military lasers are not very dependent on the atmosphere when the air is fairly clean and the laser is used at battlefield distances.

A safety analysis of the outdoor use of current military range finders and target designators indicates that these lasers can actually be used deliberately against the eyes of the enemy. However, so far as is publicly known, only one country has made such a use part of its military practice. This is the case of the use by the British of low-energy lasers to flash blind Argentinean pilots in the Falklands conflict. At present, there are no known tactical manuals which cover the deliberate use of lasers against the eyes of enemy soldiers as a weapon of warfare. What the military will do in future conflicts, which will certainly involve a mass use of

FOUR

The Laser as a Weapon

The laser beam is popularly thought of as a very powerful death ray which can be fired from a hand-held laser gun to vaporize soldiers, demolish buildings, and burn through tanks. This is an immense exaggeration which has been fostered for years by science fiction literature and movies. In reality, the laser is a suitable tool for many nonmilitary purposes, and, while it may also be turned into a deadly weapon, there are definitely limitations to what a laser can do. Although many people still consider lasers as a sort of futuristic Flash Gordon weapon, some of these science fiction depictions have become partially true. The laser really is a ray weapon, and its light rays can damage some targets in a way that appeals to the most vivid imagination. It is important to take these somewhat speculative factors into consideration when studying the psychological effects of the use of laser weapons on the battlefield. Otherwise, it will not be possible to get a complete and realistic picture of what using a laser really means to the combatants.

POSSIBLE TARGETS

A discussion of laser weapon applications outlining what laser weapons can really do must start with the destination of the laser beam—the target. The desired effect on the target ultimately decides what is needed from the laser. To a large extent, the interaction between the laser beam that is selected and the target also determines which cost-effective weapons are developed, produced, and fielded. The sensitivity of the target to laser light determines whether a low-energy or a high-energy laser is required. If the target is sufficiently sensitive to low levels of energy within a comparatively broad band of the spectrum, a cheap and cost-effective laser weapon can be designed and mass-produced. If high energy is required, the possibility of designing a usable and affordable laser weapon decreases drastically.

Right at the outset, it should be made clear that many targets exist on the battlefield which cannot be attacked successfully by any laser weapon. Furthermore, many missions cannot be fulfilled, because it is sometimes impossible to maneuver the straight laser beam around a barrier such as a line of trees to damage the target, even assuming that the target is vulnerable to laser energy. Contrary to popular belief, it is quite impossible at battlefield distances to burn holes through the thick armor of a tank or even the much thinner defenses of an armored personnel carrier or car. It is not possible for lasers to demolish fortifications or to blow up bridges and roadblocks. It is even impossible to kill soldiers instantly.

Most traditional artillery and mortars can lob shells and grenades onto enemies hiding behind mountain ridges, tree dumps, or buildings. That is not possible with a laser weapon, which is more akin to a large-caliber, high-velocity, flat-trajectory antitank gun which is dependent on a direct line of fire. It is not even possible to ricochet or bounce the beam toward the target as can sometimes be done with the shell from a direct firing gun, as this would require that some helpful person place a high-quality mirror at exactly the right spot and the right angle. Thus, the laser cannot possibly replace every weapon that is on the battlefield today. It may very well be a weapon of choice against some specific targets and can be a very useful complement to many existing weapons.

No damage will be done to the target of a laser weapon unless some of the laser beam energy is actually transferred to the target surface. The process starts as soon as the laser energy reaches the target and lasts as long as this area is illuminated. The amount of energy transferred is dependent on the type of laser, the duration of exposure to the laser beam, the interaction between the laser beam and the target, and certain environmental factors. Of all these factors, the ability of the target to absorb or reflect the energy is most crucial. A target with a high reflectivity to the laser energy may be difficult or even impossible to damage. This target property may even be used as part of a countermeasure scheme by the enemy. The laser beam can even be reflected back and cause damage to the laser weapon or its operators.

If the laser beam has a diameter of 4 inches, for example, the energy will be dispersed all over that area. The size of the laser beam at the target must then be compared with the size of the target itself. If the target is small, for example, a sensor or a human eye, it will only be affected by the laser energy in that part of the beam which actually hits the sensitive part of the target. This may mean that only a small fraction of the beam's energy actually reaches the target. That fraction may very well be enough to damage or destroy the target, but this loss of efficiency must be taken into consideration when possible laser weapons are discussed.

The main damage mechanism of laser weapons, both for antimatériel and antipersonnel use, is thermal. When the illuminated surface on the target can neither reflect nor safely absorb the beam energy, there is a rapid buildup of heat, which then melts and boils off or vaporizes the target. Even before the material in the target is heated to the melting point, it may be seriously weakened. Further damage may be caused by an indirect mechanical stress resulting from the intense heat of the laser beam or the pressure that an expanding laser spark or plasma exerts on the target. These mechanisms can deform and perforate metals. Ionization damage may result from laser-plasma-generated X rays, which can have harmful effects on many electronic components. Thermomechanical effects, due to a combination of thermal and mechanical factors (similar to what happens when glass is heated rapidly and unevenly), may cause the most severe damage. In these cases, a series of pressure waves travel through the material and may tear it apart. On a biological target such as the eye or skin, photochemical and ionization damage will be added to the thermal damage.

The only way for laser weapons to affect well-protected targets such as tanks on the battlefield is to get inside the tank through the gun sights, periscopes, or vision blocks, glass windows, and other optical pathways with the goal of blinding the crew and/or the electronic sensors. It may also be possible with sufficient energy to disable or destroy the optical pathways themselves. If the target is the unprotected eyes of the enemy infantry units fighting in the open, a low-energy weapon aligned with the sights of rifles, machine guns, and antitank weapons can be used on the battlefield up to a distance of several yards. The low-energy laser weapon may also be a way to reach human targets observing the battlefield from fortifications, foxholes, and other protected sites.

The low-energy laser weapon will be a cost-effective alternative to conventional weapons for disabling laser-sensitive electro-optics used for night fighting, surveillance, target seeking, and fire control systems. Furthermore, the low-energy laser may even be used for blinding soldiers using magnifying optics or the naked eye. A high-energy laser weapon may be used to down missiles, helicopters, and aircraft by destroying the sensors, blinding the crews, and burning holes in critical structures.

It is important to differentiate between the effects of high-energy and low-energy laser weapons. High-energy lasers may be used to melt holes through metal and plastic structures at reasonable distances, to set fire to objects, to burn a soldier's skin, and to destroy optics and electro-optical systems at long ranges. Low-energy lasers may be used to counter optical and electro-optical sensors such as thermal sights, other infrared sensors, image intensifiers, and low-light television systems, all of which are very sensitive to even an extremely low energy laser beam.

It is not possible to draw a sharp borderline between high- and low-energy weapons, although the U.S. Department of Defense defines a high-energy laser as one that has an average power output of at least 20,000 watts (or as much as 200 ordinary 100-watt bulbs put together) or a single pulse energy of at least 30,000 joules (that given off by 300 flashes from a pocket camera all at once). Some experts consider all laser applications designed to destroy material of any kind as employing high-energy laser weapons. It is true that most of these kind of attacks require a large amount of energy. However, a very low energy laser has more than enough optical punch to destroy the critical material inside a sensitive sensor or a human eye. It is always the target in question that determines the necessary properties of the laser such as the beam energy, wavelength, pulsed or continuous mode, pulse length, peak power, and so on. Since different types of targets require different laser weapons, it is of limited value to try to establish an exact energy borderline, and it must be accepted that the distinction between high-energy and low-energy laser weapons is somewhat blurred. High-energy laser weapons will generally be designed to burn and destroy targets that are not extremely sensitive to laser energy and need more force in order to be penetrated, while the less powerful weapons are designed to attack only the more sensitive targets.

If the target is an aircraft, helicopter, or missile, and the purpose is to damage the structure itself rather than a sensor, the necessary energy level will be in the region of 10,000 kilojoules per square centimeter within 0.2 seconds to get a significant effect. The energy level required to burn the skin or the outer parts of the eye is in the region of 0.2–0.4 joules per square centimeter in 10 nanoseconds for Q-switched lasers. This energy level is 25,000 to 50,000 times less than that required to damage the more resistant targets mentioned above. The structures inside the eye are even more sensitive than skin, and an energy level of only 0.5–5.0 microjoules will be sufficient to cause serious damage. This is only a tiny fraction of the energy level necessary to damage an aircraft. Sensors aboard aircraft or other machinery operating within the visible and near-infrared regions of the spectrum may be just as sensitive as the human eye, since their optics also greatly magnify any incoming laser energy. This should give an indication of the differences between high- and low-energy-level lasers. In short, only the eye and similarly sensitive sensors can be harmed by low-energy laser weapons.

The military has a choice of two main laser weapon arenas. The first includes the low-energy laser weapons designed to destroy or disable sensors, target seekers, night fighting equipment, other electro-optical devices, and even human eyes. The other alternative is the high-energy laser weapons designed to knock out targets in the air or on the ground and to destroy optics and sensors at extremely long ranges. The sensitivity of the intended target and the required range will be the two most important factors to consider when deciding whether a low-power and much cheaper laser will be sufficient or not.

ADVANTAGES AND POSSIBILITIES

One of the most remarkable properties of a laser weapon is its speed. The "laser bullet" moves at the speed of light, 187,281 miles per second. By comparison, the projectile of the usual tank gun has an estimated speed of about a mile per second. This means that when a tank crew fires at another tank 2 miles away which is moving at a speed of 23 miles an hour (11 yards per second), it will have moved 22 yards before the projectile arrives. This is, of course, a problem for the tank gunner, who has to calculate the true speed and direction of motion of the target. Having done so, the gunner must lead or aim ahead of his target at the point where the tank will be when the projectile arrives. A high hit probability can only be achieved with the assistance of a fire control system that calculates this factor and provides the correct aiming point. Then, all the gunner has to do is press the button and wait for the result. It is of interest to note that one of the basic components in most such fire control systems is a laser range finder.

The problem of determining the correct lead for a moving target (that is, the distance that the gunner must aim in front of it) is even more difficult when the target is traveling rapidly, as in the case of an aircraft. The usual speed for a military aircraft during an attack may be 300–330 yards per second or 610–740 miles per hour. If an aircraft moving at 330 yards per second is shot at by an antiaircraft gun at a distance of 3 miles with a projectile whose speed is 1,375 yards per second, the lead has to be at least 1,260 yards for a side view of the airplane. If the aircraft is not moving in a straight line and has a more or less oblique course compared to the projectile, the problem is much more difficult. This is why it is necessary to have a very complex fire control system to solve this three-dimensional problem. Yet, despite all the extra assistance from the fire control system, the hit probability may still be rather low. In comparison, the beam from a laser weapon will travel at the speed of light, and the aircraft will move only two yards before the laser beam hits it. The tank will move even less. These small movements are without any importance whatsoever in relation to the weapon's immediate effect on the target. If the target is resistant, and it is necessary to attack the same spot for some time in order to get the desired effect, the problem becomes a little more complicated. However, it should be remembered that the laser gunner has an almost zero time-of-flight weapon, and it is certainly not possible for the target to take any evasive action to avoid the beam or to activate any protective countermeasures before it is hit the first time.

When a tank gunner is fighting on the battlefield, he often has more than one target to engage at the same time. It is vital to him that he can destroy or neutralize these targets in rapid succession without being hit himself. This sequence usually takes a relatively long time to complete. The gunner has to use the fire control equipment to aim, then press the button to fire, wait for the projectile to reach the target, assess the result, sometimes fire a second round, and only then engage the next target. However, if the gun is replaced by a laser weapon, only an extremely short time is needed for reaction, target engagement, and target changing. There is no shooting time, and the laser gunner has only to press the button to get the next "round" off. Such a laser weapon may engage and destroy many more targets in a given time than a conventional gun. The differences are, of course, measured in a few seconds only, but these few seconds may decide who wins and survives. This line of reasoning, of course, assumes that the laser weapon can really damage the target sufficiently. Otherwise, the conventional gun or missile will obviously be the first weapon of choice.

Rifles, guns, mortars, and related conventional weapons all have to get their ammunition directly to the target in order to be effective. Large variations exist between the different conventional weapons in this respect. Weapons such as tank guns, for example, have a rather flat trajectory, with only a yard drop below the straight line between the weapon and the target. This is usually the case with weapons using a direct line of sight and having a high-speed projectile (or muzzle velocity). Weapons that are aimed indirectly, such as howitzers and mortars, have an arc trajectory or line of projectile flight which, in some cases, is as high above the ground as the distance to the target. Because they have such an arc trajectory, they can reach targets behind hills, woods, and other terrain features. This is an impossible task for a laser weapon, which really cannot bend its beam around any hindering objects and thus can only be compared to weapons with a very flat trajectory. However, even for a tank gun with a comparatively flat trajectory, calculations are necessary to compensate for the motion of the target during the projectile's time of flight. This is not necessary with a laser weapon, because it has a perfectly straight line of fire which is identical to or parallel with the line of sight and a beam which can instantaneously reach a target before it moves away.

Some weapons such as cluster bombs, artillery mortars, and machine guns cover a wide area but have a rather low hit probability against small, single targets. In many situations and especially when the soldiers are under severe combat stress, it is easy to miss the target by spraying a weapon's ammunition haphazardly over a wide area. Laser weapons give soldiers the ability to position the beam on an individual target with a high degree of accuracy and with no risk to bystanders, even if they are very close to the target. Thus, the laser is truly a weapon that may be used selectively if required. However, this is not true if a laser weapon is designed to use a scanning beam covering large areas, which might be necessary in the heat of battle. In such cases, everybody will be hit including both friendly troops and civilians.

Another significant characteristic of low-energy laser weapons is their ability to operate in almost complete silence. This can be contrasted with the very noisy performance of rifles, machine guns, and tank guns. The position of these lasers will not be given away by their noise when they are fired, thus providing an important advantage to the laser gunner. High-energy lasers are very different in this respect, since the massive flow of energy pouring through the weapon is very noisy; indeed, the noise is comparable to that of conventional field artillery. Even the passage of the high-energy beam through the atmosphere may be heard as dull thunder or crackling.

The radiation from lasers that operate outside the visible region of the spectrum is invisible to the human eye, which means that the location of such lasers cannot be pinpointed by visual observation. Laser beams in the visible part of the spectrum can be observed outside the beam path itself only when atmospheric contaminants in the beam reflect light out of it. Many people have seen this effect at laser light shows. This may be a problem to laser gunners using lasers emitting in the visible part of the spectrum, especially at night, as the laser beam may give away their position. However, if the designer of a laser weapon uses a pulsed laser, the possibility of the beam being seen off axis may be substantially reduced. If a laser weapon is operating in the infrared or the ultraviolet, special instruments will be needed for detection of the laser beam.

A major factor that favors the use of laser weapons, and especially the low-energy lasers, is that they do not need the huge and expensive ammunition logistic system associated with the conventional weapon systems of today. Ammunition is tremendously heavy and bulky and must always be stored, transported, and used with some risk of premature detonation. The low-energy laser may only require a battery as "ammunition," and, with such an energy source, it will be capable of firing tens or hundreds of shots. This alone will have a great impact on the cost-effectiveness of low-energy laser weapons. On the other hand, a high-energy laser will burn a lot of fuel, which will have to be produced, stored, and transported to the battlefield. A typical CO_2 anti-aircraft laser weapon may use 5–10 tons of fuel to fire 50 shots. This constitutes a major logistical problem but not an overwhelming one compared to the logistical problems of other air defense systems with a similar range of 6 miles.

The cost of any weapon is always an issue of high concern for the military. The original cost of a low-energy laser weapon may not be too high compared with the costs of more conventional battlefield weapons. A very simple anti-eye laser weapon will not cost more than an ordinary hand-held laser range finder and, in fact, may cost much less. A more complicated low-energy antisensor laser weapon may cost approximately two to five times more than a laser target designator if produced in large quantities. If a low-energy laser weapon is made more sophisticated by adding various high-technology accessories, it can become much more expensive, but, at the same time, it will be able to destroy much more sophisticated targets. In short, a low-energy laser weapon can be made cost-effective in a fairly easy manner, as it may destroy or block such valuable targets as highly delicate sensors and human eyes. On the other hand, a high-energy laser weapon will certainly be quite expensive. Already, billions of dollars have been spent on developing such weapons without any notable successes. It is even still doubtful if it will be possible to develop, manufacture, and put cost-effective high-energy lasers into the field before the turn of the century. The major obstacle for high-energy laser weapons at present is not the exorbitant cost but rather the technological barriers that need to be overcome.

The size of a laser weapon is an advantage only in the case of low-energy lasers. A high-energy laser designed to shoot down aircraft and missiles may be the size of a truck or tank, while a simple low-energy anti-eye laser may be as small as a pair of binoculars. The existence of various laser weapons within most armed forces today means that there are a lot of officers and enlisted men who are already accustomed to directing beams against different targets. Laser technology is not new to combat units, which means that maintenance, battery supply, and risks to personnel are already familiar problems. It will not be very difficult from a technical standpoint to field and use low-energy laser weapons on a large scale in the future, although, of course, there will be other issues to deal with related to tactics and even ethics.

From the manufacturing or supply standpoint, many countries all over the world have some kind of laser industry of their own. It will certainly be possible in most of these countries to produce simple low-energy military lasers, range finders, etc. More complicated low-energy lasers useful for antisensor weapons require a laser industry with a high level of laser technology for research, development, and production. Such facilities only exist in countries with a more advanced electronics industry. High-energy weapons will be researched, developed, and fielded only in a few highly developed countries that possess enough resources in manpower and have an adequate industrial economy. However, when discussing the future of laser weapons, it is important to keep in mind that uncomplicated low-energy weapons may be mass-produced, not only in all industrialized countries but even in many Third World countries as well.

DISADVANTAGES AND DIFFICULTIES

Obviously, the laser is not capable of damaging or destroying every kind of target on the battlefield. In the near future, the laser will only be used against targets which are very sensitive to laser energy such as sensors, electro-optical devices, and, possibly, using a longer term perspective, human eyes. Beyond the year 2000, the laser may be used to destroy tougher targets such as aircraft and missiles. Laser weapons will normally not be used completely on their own but will be one of many weapons in the military's arsenal working together on the conventional battlefield.

When gunners of rifles, machine guns, or other sorts of barreled weapons or missiles are firing at a target, in most cases the effect on the target can immediately be seen by the gunners' own eyes or by some optical instrument. If the target is not hit, the gunner can determine by how much it was missed, make the appropriate corrections for distance and direction, and fire again. This is not the case with a low-energy laser weapon. If the laser gunner uses an invisible laser beam or even a beam in the visible part of the spectrum that is difficult to identify, direct hits will not be registered. Furthermore, it is not easy to judge what effect the beam has had on a target. In the case of a miss, the gunner even has difficulty determining by how much and in what direction the beam was misdirected. The outcome of firing may be easier to ascertain with a high-energy laser weapon, which will have a more obvious and violent effect on the target, but it will still not be as evident as with more conventional weapons.

One of the main difficulties with laser weapons is their dependence on conditions within the atmosphere. Even on what seems like a very clear day, the air may appear to be much more transparent to the laser beam than is actually the case. These problems will certainly become worse on the battlefield, where the air is often filled with smoke and dust. Furthermore, rain and fog can have a devastating effect on some lasers, reducing their effectiveness to almost nothing. The designer of laser weapons has to take all of these factors into consideration; the use of the appropriate laser characteristics, the proper accessories, and enough energy can circumvent most atmospheric problems. Also, intensive training is necessary to provide the laser operator with a thorough knowledge of all the problems involved in operating a laser weapon.

The main effects of the atmosphere on the laser beam are absorption, scattering, turbulence, spark generation, and what is called thermal blooming. The theory behind laser beam propagation through the atmosphere is really quite complex, and extensive research has resulted in many comprehensive books and a plethora of articles on the subject. Most of the research has been done in connection with military laser programs, which gives one indication of the extent to which military funds are supporting the development of laser technology.

Turbulence in the air may bend and defocus the beam. The density of the air is always fluctuating randomly due to winds, differences in temperature, local topology, and many other factors. This leads to spatial fluctuations in optical properties in the air. Turbulence causes deformation of the beam on its way from the laser optics to the target. If the firing range is long, the beam cross section may deviate considerably from the ideal circular shape. Inhomogeneities in the atmosphere along the beam path may act as lenses to give hot spots in the beam. Hot spots are localized areas of the beam where the intensity of the beam is much greater than the average due to localized focusing effects. Fog, rain, smoke, or other obscuring haze weakens and scatters the beam but does not cause hot spots, and, of course, if the beam is absorbed, it will not reach or destroy its target.

The efficiency within the atmosphere of high-energy laser weapons is reduced by several more factors. One of the most important of these is a phenomenon called thermal blooming. It is caused by a succession of sparks or plasmas created by the beam's energy which heat the air within the beam. These small heated areas are difficult for the energy to penetrate, and that is why they cause the beam to diverge or "bloom," reducing its efficiency at long ranges. As the power of the laser beam is diffused, it becomes increasingly difficult to damage the target. The problem of thermal blooming becomes even worse if it is necessary to track a moving target and, thus, move the laser beam through the air. This will be the case for almost all high-energy laser weapon applications where it is necessary to heat a spot on the target for some time. The beam bends in the direction of the wind and spreads out, producing a focal spot that is longer in the direction perpendicular to the wind. Thermal blooming is an effect which increases rapidly as the beam power increases, and, in the cases when the beam power exceeds a critical value, the intensity on the target decreases rapidly. It will certainly be necessary to compensate for these atmospheric influences if high-energy laser weapons that are effective at long ranges are to become a reality. Some methods of minimizing thermal blooming will be described in later chapters, but it should be understood from the start that there is still some uncertainty about whether or not it is possible at all to solve the problem satisfactorily. If these atmospheric influences are not compensated for, the intensity on the target from a high-energy CO_2 laser at a range of a few miles may be reduced to values of a few hundred watts per square inch. The values from a high-energy DF laser will be only slightly better.

Another effect caused by the passage of the high-energy beam through the air is stimulated Raman scattering, which changes the original wavelength of the beam to another or several other wavelengths. This is a process that may repeat itself and lead to the beam losing much of its efficiency. The research done in this field is very incomplete, and there seems to be a general lack of information. Some researchers have shown that this effect does not occur at all wavelengths so that perhaps a tunable free-electron laser (FEL) could solve the problem. If further research reveals a group of wavelengths that have good transmission capabilities, the FEL could easily be tuned to it.

The negative effect of atmospheric conditions on high-energy laser weapons means that the effective range of the beam will be considerably shortened even in seemingly good weather. Under more severe conditions such as rain, fog, or snow, effective laser action in air defense may be almost impossible. An air defense high-energy laser weapon cannot reach aircraft or missiles through thick clouds, and, furthermore, it is difficult to quantify the limitations due to the pollution of the atmosphere that can be expected to occur on the battlefield. However, these atmospheric disturbances are not too important when it comes to close combat situations on the ground. The average fighting distance may be less than a mile even in a tank-versus-tank engagement. It is perfectly possible to design lasers operating against sensors and human eyes with an energy output that is sufficient to cope with most atmospheric situations up to two miles.

The single and direct line for aiming and firing that character-izes the laser weapon concept is a great advantage, but, at the same time, it can also represent a disadvantage. When the gunner fires a laser weapon against the enemy soldier, the laser's position may be given away by the beam if the enemy has a laser detector or if the beam is visible, and thus the gunner may become a target himself. This is, of course, the problem with all direct-line-of-fire weapons; however, the laser still has the advantage of silence and possibly invisibility. As has been said previously, the direct line of fire also means that it is impossible to use laser weapons for indirect fire.

The high-energy laser weapon concepts that have been dis-cussed so far all involve far more difficulties than the low-energy applications. Another major drawback of high-energy lasers is their high fuel consumption in addition to the inherent noise and bulk of the machinery. When these factors are added to the difficulties with the beam transmission through difficult atmo-spheric conditions and thermal blooming, it is clear that the solutions will obviously involve very costly and complicated tech-nologies. These added difficulties with high-energy lasers empha-size that low-energy laser weapons offer a far more realistic and certainly cheaper alternative for the time being.

A large-scale deployment of various laser weapons on future battlefields will have a strong influence on how combat units will have to fight. A laser weapon doctrine has to be worked out, both for the offensive use of these weapons and the defense against hostile lasers. Such a doctrine should cover all aspects from army group operations down to the role of the individual soldier, airman, and sailor. The implementation of such a comprehensive doctrine will take time, require a lot of training, and certainly cost a great deal of money. It is necessary to make very precise arrangements to ensure that combat units know how to cope with the laser threat; otherwise, they are risking a very nasty surprise in actual combat. Optimally, laser weapons should be combined with conventional weapons in a way that will produce as high an efficiency rate as possible in combat.

THE MAJOR LASER WEAPON CONCEPTS

There is generally more than one laser weapon alternative for each proposed laser weapon mission on the battlefield. It is quite possible to vary the laser properties and energy level, the tracking system, and the fire control equipment according to the military requirements for each specific mission. Environmental influences will also have a very strong impact on the final choice of laser weapon applications. For example, a hydrogen fluoride laser is not the best choice for long-range use within the atmosphere, because its wavelength is strongly absorbed by the atmosphere. Every laser weapon that is designed to operate within the atmosphere over any great range, whether ground-, sea-, or air-based, must use wavelengths at which the atmospheric absorption and scattering are as small as possible.

Laser weapons may be used within an army's air defense against aircraft, helicopters, and missiles. The desired effect on the target may be either to burn holes and destroy key structures, to blind or trick the sensors, or to blind the crews temporarily or permanently. The high-energy air defense laser may use all three effects at the same time if the target is within reach of the main effect of the laser. At longer distances, only the antisensor and anti-eye capability will be possible. The low-energy air defense laser will use enough energy to be effective against sensors and eyes. It is also possible to field a laser with the main purpose of blinding or flash blinding the crews. Flash blinding will be most effective in the dark when the eye is dark adapted and much more sensitive to overload by bright flashes.

The cost-effectiveness of low-energy defense laser weapons will be very substantial if it is realistically possible to knock out a very expensive enemy aircraft or helicopter with a relatively simple and cheap laser. The cost-effectiveness of the much more complicated and certainly very expensive high-energy laser weapons is more questionable. Such weapons may be worthwhile only when they protect very costly objects such as air bases, ships, and high-level command posts.

Both high- and low-energy antisensor laser weapons will be used primarily against optical and electro-optical equipment. The required effect on the target will be to destroy, fool, or blind the sensors. This application has already been mentioned in connection with the high-energy air defense weapons, and it is certainly also a valid alternative on the ground. Low-energy laser weapons may be used on the battlefield against night fighting equipment, tank fire control systems, missile sights, and other similar sensor systems. It is also possible that a high-energy laser might be a suitable weapon against the vision blocks, optics, and sensors on modern tanks and other fighting vehicles. If this turns out to be a cost-effective solution, it will probably add a longer range to present applications that employ low-energy systems.

One very interesting laser weapon concept is to try to counter sea-skimming missiles while they are approaching the target ship. This may be accomplished with an antisensor laser or with a high-energy air defense weapon or a combination of both. Some missiles and other weapons such as smart bombs home into the target with the help of low-light television or an infrared seeker— systems which are very sensitive to laser action and, therefore, can be easily manipulated by a laser weapon. It may even be possible to field an airborne laser weapon to counter some enemy air-to-air missiles. The problem is that the weight and size of the laser and its tracking system may be too much for a combat aircraft to carry. Another interesting concept is to direct the laser beam from the aircraft against the sensors and sights of the air defense sites. Some ground-to-air missiles use a beam riding technique, and it should be possible to fire a laser weapon from the threatened aircraft along the laser beam that is attacking it back to the missile operator or the missile itself.

The most spectacular and horrifying use of the laser weapon may be in the antipersonnel role, where the main target is the human eye. The desired effect may be to flash blind the adversary for shorter or longer periods of time or even to permanently blind. The most obvious targets are the eyes behind the magnifying optics in armored vehicles, missile systems, artillery control systems, and many other places on the battlefield. The use of magnifying optics means that the effective range of the laser weapon will be substantially increased, as the effect of the incoming laser light is increased by the magnifying optics, but even ordinary infantrymen can be a target for anti-eye laser weapons. It is surely possible to design and field cheap laser weapons that can be added to ordinary small arms and used in close combat at distances of up to 1,100 yards. The cost of such a weapon drops with mass production, and the broad-scale fielding of this type of weapon would change not only the tactics and battle doctrine of combat but would even affect the requirements for medical services, because of the large-scale casualties that would be expected from such engagements.

High-energy laser weapons could also be used as very sophisticated and long-range flamethrowers. It would be easy to set fire to trees, wooden buildings, uniforms, and other flammable objects at long distances, causing widespread fires and a devastating psychological effect on the enemy.

As will be discussed in the following chapters, the widespread use of the military laser weapon, both against equipment and personnel, is rapidly approaching becoming a reality, but there are many limitations.

FIVE

High-Energy Laser (HEL) Weapons

Laser weapon projects have always been shrouded by very tight security. In spite of this, it is possible to follow the general lines, at least, of the high-energy laser (HEL) weapon research field through the open literature. This is especially the case with the development of different laser projects in the United States. Most of the following discussion and evaluation refers to well-publicized U.S. programs. However, this should not lead anyone to believe that the laser weapon business is mainly an American affair. Much work is going on not only in other Western countries but also in the Eastern bloc. Even though our information from the Russian part of the former Soviet Union on the state of their laser technology is very sparse, there can be no doubt that they are working very hard to actualize laser weapon ideas. According to David Isby, author of *Weapons and Tactics of the Soviet Army*, in 1988 the Soviets were at about the same level as, or even more advanced than, the West in the development of offensive laser weapons. He has also stated that the Soviets had begun practicing weapon applications with a variety of laser technologies that are still in the realm of pure scientific research in the United States.

The efforts to develop and field high-energy laser weapons were initiated as soon as the first lasers capable of delivering high energy were introduced in the late sixties. The gas dynamic carbon dioxide (CO_2) laser was the earliest truly promising high-energy laser concept and was developed in the United States by AVCO Everett in 1968. This was soon followed by the hydrogen fluoride (HF) and the deuterium fluoride (DF) chemical lasers developed by the United Technology Research Center (UTRC) in 1969. In the early 1970s, all three military services in the United States started research programs largely based on theoretical considerations aimed at investigating the vulnerability of relevant military targets to high-energy lasers.

One of the most important events fostering the development of HEL weapon technology was the Strategic Defense Initiative (SDI) established by the U.S. President in March 1983. The aim of this program is to use different weapons (many are beam weapons, including various laser types) which will kill incoming intercontinental missiles and warheads mainly outside of the Earth's atmosphere. High-energy lasers represent one of several possible classes of weapons that have been intensively discussed and researched for use in this program. The lack of atmospheric influences is, of course, a significant advantage for all high-energy lasers at the enormous ranges that are involved. Although the conditions in space and within this entire strategic warfare concept are quite different from those on the conventional battlefield, the enormous resources in money and manpower that have been allocated to the development of laser weapon technology within the SDI program have certainly speeded up the progress of conventional battlefield laser weapon programs.

Despite all of the SDI efforts to date, no high-energy laser weapons have even been fielded in space or on the ground. This gives an indication of the magnitude of the difficulties involved and indicates that, for the near future at least, large-scale fielding of HEL weapons designed to destroy relatively hard structures such as aircraft seems unlikely.

HEL TARGETS

The main use of HEL weapons will be for air defense, and vigorous efforts have been made by some countries to investigate seriously the use of HEL weapons for this purpose. Defense staff military planners, scientists, and engineers at industrial research institutes worldwide have worked hard at trying to design and field HEL weapons that will meet the growing threat from increasingly sophisticated attack aircraft, armed helicopters, and a growing number of different missiles, including sea skimmers. In theory, the military requirements are quite simple; the HEL weapon must be able to destroy the airborne targets at night as well as in bad weather before they deliver their load of munitions on a protected facility. If the aircraft release their payload outside the range of the HEL weapon, then it must also be capable of destroying the incoming munitions before they can accomplish their mission.

The air defense environment is usually complicated by a high degree of atmospheric pollution, yet, despite this problem, a very high standard of performance will still be required from any laser weapon system. When important targets need to be protected, it is necessary to take into account the enemy's probable use of a large number of attacking aircraft or helicopters equipped with the most modern weapons. In most cases, this could mean four to eight aircraft attacking simultaneously from several directions. Modern technology allows an aircraft to fly toward the target area close to the ground and deliver its munitions from a very low altitude. In some cases, long-range weapons will be used whose missiles can be launched at the final target from the attacking aircraft well outside the range of the defending laser weapons. Of course, the greater the distance from the target that the launching takes place, the lower is the probability of a successful hit.

Aircraft, helicopters, and missiles are becoming increasingly faster, more intelligent, and much more versatile, which means that all types of air attackers will have to be destroyed or countered before they have time to use their weapons. It is no longer sufficient to neutralize the majority of the attackers; it is now necessary to shoot down or counter almost every one of them. This will become even more difficult as future combat aircraft and helicopters will be harder both to detect by radar and to hit, as they will be protected to some extent against air defense weapons. The attacking enemy aircraft will also be supported by airborne electronic countermeasures, will presumably be well informed, and will be guided from airborne command posts. The attackers may even use smart, almost jamproof missiles or remotely piloted vehicles.

The air defense of today is composed of a combination of interceptor and fighter aircraft, antiaircraft guns, and missiles directed by trained command control and intelligence systems. Even if, in most situations, the combined effect of all air defense units is sufficient to cope with the present threat, it will not be so in the future. When an important target is to be protected properly, it will be necessary to stop virtually all attacking airborne weapons.

Although modern antiaircraft guns have a high rate of fire and use high-speed projectiles or, at least, ones that leave the gun barrel with a high muzzle velocity, they still need the advance knowledge of the target path or trajectory. The presence of electronic countermeasures aboard enemy aircraft in many cases will give an unacceptable kill probability. Furthermore, an artillery system often needs several hits or near misses to down a single target depending on the caliber of the gun in question.

Also, contrary to what is sometimes believed, missile weapon systems do not have a 90% or more hit probability. Even if antiaircraft missiles are properly handled, the very hard and sometimes unpredictable realities of the battlefield have been shown to necessitate devastating revisions of peacetime data and calculations of weapon efficiency. When the enemy countermeasures and evasive actions are added to the quite normal difficulties resulting from a very stressful and life-threatening situation, air defense missile units are often very satisfied if the kill probability exceeds 50% for each missile fired. The true figure is usually less than that. A missile system has a relatively long reaction time from target detection until missile launch, often more than five seconds. If these crucial seconds are added to the five to ten seconds it will take the missile to fly to the target, the possibility of engaging any given target successfully becomes somewhat limited. If the target moves at a very low altitude—"tree skimming"—with a speed of 300 yards per second, it will cover at least 1.5 miles before the missile can possibly hit it. This may not be too problematic if the number of targets is the same or nearly the same as the number of guns or missiles and the enemy is flying at an altitude that makes it possible to engage him. If there are multiple targets for every gun and missile, however, there is a good possibility that a substantial number of them will get through. Any enemy will certainly be aware of these facts and will try to attack important targets with as many weapons and aircraft as possible. The most difficult and extreme case will be when the attacker can launch multiple missiles or bombs at extreme ranges. Thus, it is already very difficult and will become even more complicated in the future to defend high-value targets effectively against airborne attacks with conventional missile and gun system technology.

Some military scientists and staff members have advocated the introduction of the HEL air defense weapons on the battlefield as the only solution to these problems. According to one of the individuals involved in the present development of a German air defense HEL weapon, the following essential military require- ments must be fulfilled in order for a future HEL system to cope with even a present-day threat. The air defense laser must have multiple target detection and tracking ability with a target detec- tion time of less than 1.5 seconds. The aiming time should be less than 0.5 seconds for the first target and 0.1 seconds for each additional target in a group. In addition to these extremely short reaction times, an additional requirement is that there be a sufficient amount of fuel to allow ten or more laser shots to be fired within very short time limits. Finally, the tracking and fire control systems must provide for a very high hit probability.

Approaching missiles, due to their small size, high speed, and possible large numbers, are usually more difficult targets than aircraft or helicopters. Only if the missile is dependent upon and equipped with a sensor that is sensitive to laser light does it become easy to disable before it reaches the target. One very interesting application of the HEL weapons is to protect ships against sea-skimming missiles, which have a long flat flight path. That such missiles pose a very serious threat to any surface warship has been recently demonstrated in the Falkland Islands conflict, where EXOCET sea-skimming missiles launched by the Argentineans sank both a destroyer, the HMS *Sheffield*, and a commercial container ship, the HMS *Atlantic Conveyor*.

One of the advantages of using a large ship as a base for an HEL weapon is the possibility of using the ship's main drive engines as an electrical generator to power the laser. This means that there will not be any shortage of ammunition and that it will be possible to fire many laser shots within a very short time interval. The laser weapon with its direct line of sight in both elevation and azimuth, its almost zero time of flight, which also eliminates the need for a lead, and an almost unlimited supply of energy seems to be an ideal weapon for the protection of valuable ships. However, there are still many problems; the humid atmosphere and often severe weather conditions surrounding vessels at sea may be the most difficult issues to solve.

If the primary goal for a given HEL weapon is not to destroy the target itself but rather to attack battlefield sensors, night fighting equipment, fire control systems, and other electro-optical devices, less energy will be required, and, thus, it will be much easier to cripple or blind the target. An antisensor HEL weapon may thus be used at much longer ranges than a laser weapon designed to burn holes and destroy hard targets, or it can be used at the same range with much less energy. Since most aircraft and helicopters are equipped with several sensor systems, which are a necessary part of any attack against targets on the battlefield, in the air, and at sea, the attack may very well be neutralized, for some time at least, just by destroying or blinding the sensors. This indirect application of the HEL may be more cost-effective than more ambitious attempts to shoot down the aircraft itself.

The HEL weapon can also be given to combat units as a very efficient flamethrower, since it can set fire to flammable objects on the battlefield at very long ranges. The enemy soldiers may be burned out of buildings, grassy areas, brush, and forests. Human beings are afraid of fire, and this application may very well be used as a psychological weapon to terrify the enemy infantry. The risk of setting a soldier's uniform on fire may also have a devastating effect upon his morale and will to go on fighting. However, the high economic cost of this application will almost certainly limit its use to a few very strategic enemy positions.

ENERGY LEVELS AT THE TARGET

One of the basic questions facing the laser weapon designer is what energy level must be absorbed by the target in order to get the desired result. The absorbed energy (E) is some fraction (A) of the product of the power density or intensity (I) present in the laser beam and the emission duration (t). E is measured in energy units, joules (J) or watt seconds per area, usually expressed in square centimeters, I in power units, watts (W) per square centimeter, and the time in seconds in the following equation:

$$E = A(I \times t)$$

This means that if the emission duration is required to be short, as it would be in the engagement of multiple targets, the power density has to be as high as possible. The power density is calculated as the beam power divided by the size of the "beamed" area, which means that a high beam power and a small surface area will give a high power density. How much of the laser power will finally be absorbed by the target in the affected surface area will determine what destructive effect will be achieved. The laser power goes from the laser to the target, suffers transmission losses in the optical system and the atmosphere, and has a further loss when some of the power is reflected from the target surface. The absorbed power is normally no more than 20–60% of the original emitted laser power.

The effectiveness of a laser beam in causing mechanical damage is, thus, dependent on beam power, pulse duration, wavelength, air pressure, the material, and the finish of the target surface. For example, a painted area has a considerably increased energy absorption when compared to an unpainted aluminum plate. The absorption varies widely between different materials and at different wavelengths. The absorption of a ruby laser at 694 nanometers is 11% for aluminum, 35% for light-colored human skin, and 20% for white paint. The corresponding figures for a CO_2 laser at 10,600 nanometers are 1.9, 95, and 90%. This also indicates that one way to counter a HEL weapon is to choose a very reflective material for the target surface. On the other hand, longer wavelengths emitted by the laser can reduce the effects of highly reflective materials and increase the absorption. Every factor in this very difficult pattern combines to determine the degree of target destruction as well as the final energy level that will be needed to produce the desired effect.

It is obvious that the level of energy required to destroy a target varies considerably depending on the circumstances. Therefore, it is not surprising that the required energy level figures quoted in the open literature also show rather large variations. In spite of this, some numbers may be given which indicate the general range of energy levels.

An aircraft, helicopter, or missile could be hit with an HEL weapon in many different ways that in the end would nullify it. Fuel tanks could be ruptured, or the fuel itself could be caused to explode. Windshields could be shattered, and parts of the control surfaces such as elevators or rudders could be destroyed or disturbed enough to make it impossible to continue flighting. The rotor head of a helicopter or the wing of an airplane or missile could be made to fail, resulting in a crash. Sensors, radars, and other navigation aids could be destroyed; if this destruction occurs during a sensitive and crucial moment in the last phase of an attack, it could result in a crash or an aborted mission. Also, in some situations, an HEL weapon could even explode the ammunition carried by an airborne attacker.

To punch through the metal skin of an airplane requires about 700 joules per square centimeter, although it should be noted that a hole burned in the skin of an airplane may not be sufficient to destroy it in the air or even to make it crash. A more realistic energy level to disable an aircraft may be five to ten times higher, which means that a successful HEL weapon will have to be able to deliver at least 5,000–10,000 joules per square centimeter on the target.

Optical sensors and radomes (plastic radar domes) are much easier to damage; no more than 10 joules per square centimeter needs to be delivered directly on the target. Furthermore, if the laser wavelength is within the sensitive wavelength region of the sensor in question, the energy needed could be extremely low. If the HEL weapon is used as an antipersonnel weapon, that is, as a long-range flamethrower, the energy necessary to burn exposed skin is merely 15 joules per square centimeter, and damage to the cornea, the clear window into the eye, requires only 1 joule per square centimeter.

An air defense HEL weapon designed to shoot down airplanes, helicopters, and missiles successfully must have the ability to keep a very powerful beam at one point on the target for a long enough time to deliver at least 5,000 joules per square centimeter. This requires a laser in the megawatt range. If the shot is to be successful, it must be directed to a certain part of the target that is limited in size and very sensitive and then kept there until the desired effect is reached. Thus, the laser beam must track and follow a target if any great length of time is needed to achieve the desired effect.

Many parts of an aircraft or helicopter are highly resistant to an HEL weapon, but there are still enough thin-skin parts and sensitive areas to produce a devastating effect or destruction if hit precisely. On the other hand, it is obvious that at battlefield ranges even an extremely high energy laser weapon cannot penetrate the heavy armor on a tank or other armored vehicles and thus an HEL weapon is of no use for destroying resistant ground targets in the battlefield. However, sensors, optics, and related devices are still valid targets wherever they appear on the battlefield, even in a tank.

CHARACTERISTICS OF HIGH-ENERGY LASER (HEL) SUITABLE FOR USE IN WEAPON SYSTEMS

The rapid development of laser technology has led to hundreds of different kinds of lasers, but only very few of them may be scaled up into the high-energy field. Carbon dioxide (CO_2) lasers are the most obvious possibilities for use in HEL weapon applications. Carbon monoxide (CO), hydrogen fluoride (HF), deuterium fluoride (DF), and iodine:oxygen ($I_2{:}O_2$), as well as the free-electron (FEL) and X-ray lasers, along with argon fluoride, xenon fluoride, and many other types of ultraviolet excimer lasers are also candidates. HEL weapons produce a huge internal amount of heat, and prolonged operation at very high powers requires an effective system for the disposal of this wasted heat. In a gas laser, the high fuel flow serves to remove the excess heat, as the fuel is warmed by the laser reaction chamber and, in the process, cools the laser. Most high-energy lasers now under development are gas lasers working in this way. Such a laser will sound and, to some extent, look like a jet engine. Indeed, in the HEL field today, only the X-ray and free-electron lasers are not gas flow systems.

The laser in an HEL weapon system has to emit an average beam power of several megawatts during the required exposure time. This power level is two or three orders of magnitude higher than that used by the most powerful industrial processing lasers.

This power requirement together with the adverse environment in outdoor use under battlefield conditions makes the design task even tougher. When all aspects of the HEL weapon problem have been considered, very few real possibilities remain.

The gas dynamic CO_2 laser is one of the few lasers that shows promise in the HEL weapons field. The fuel may be a common hydrocarbon, for example, benzene (C_6H_6), which is burned together with an oxidizer such as nitrous oxide (N_2O). The fuel can easily be carried in liquid tanks, and the waste gas mixture is nontoxic. The wavelength is between 9,350 and 10,600 nanometers, and, theoretically at least, it is possible to have an average beam power of over five megawatts. The technology for operating this laser is rather well known and highly developed. Of course, there are some disadvantages. The very high output gas temperature has a bright IR signature. That is, the temperature is easily detected by enemy sensors. Also, there is a high risk of causing fire in the surrounding environment because of the hot exhaust gases. This laser will be rather bulky, of comparable size to a battlefield tank. As will be described later, much research is going on to solve the technological problems of high-pressure combustion and adverse changes in beam quality while the atmosphere is being traversed. The use of the gas dynamic CO_2 laser seems to be one of the more realistic HEL weapon concepts, and this type of laser has already been used in quite a few military developmental programs but as yet has not become an operational field weapon.

The CO laser operates at several wavelengths within the spectral range between 4,700 and 6,200 nanometers, but poor atmospheric transmission, mainly as a result of water vapor absorption, effectively limits its usefulness to wavelengths shorter than 5,000 nanometers. Electrically excited versions of both the CO_2 and CO lasers are not as promising as the gas dynamic versions. Both require a relatively large energy supply with a poor overall efficiency. Even so, electrically excited versions have been tested in some experimental HEL weapons.

The HF laser, operating in the spectral range between 2,500 and 3,000 nanometers, is not the best laser to use within the atmosphere because of very strong atmospheric absorption in that part of the spectrum, but it is relatively cheap and has a simple design. It is probably more useful in military space programs. The DF laser with the same design uses a wavelength of 3,800 nanometers, where the atmospheric transmission is fairly good. The DF technology is mature, and the laser has a low infrared signature and high efficiency with sufficiently good beam quality. In spite of the high price of deuterium and difficulties with the chemical pump technology, the DF laser is still a realistic option for a battlefield laser weapon.

The chemical $I_2:O_2$ laser is a new and still somewhat unknown high-energy system. A chemical reaction excites oxygen molecules, which transfer their energy to iodine atoms. The wavelength is 1,300 nanometers, which is transmitted rather well through the atmosphere. There is a developmental program for a 50,000-watt iodine laser in the United States, and several reports indicate the construction of Soviet iodine lasers. The information available gives no indication of the future prospects of this laser.

The free-electron laser (FEL) has the potential of generating very high powers and is, therefore, considered very suitable for use as a laser weapon. The SDI program proposes to have an FEL operating on the top of a high mountain directing its beam toward an orbiting relay mirror which will then deliver the energy to a target in space. The big advantage of using the FEL as a battlefield weapon is the capability of selecting a wavelength that is appropriate to the military target requirements and optimizing atmospheric transmittance. However, the possibility of scaling down the present FEL size to one that is useful and practical on the conventional battlefield still seems far away. One of the main centers of research on FEL weapons is located at the Los Alamos National Laboratory, where, in 1989, an existing FEL was adapted for tests within the SDI program. A photoinjector device replaced the cumbersome and expensive electron gun previously used for the creation of the laser beam. The electron gun was truck-sized, while the photoinjector is close to the size of a bread box. Furthermore, the laser beam may be 100 times brighter than those of FELs using an electron gun. However, even the rebuilt FEL with its electron wiggler, all high-voltage accelerators, and the photoinjector will still be a very large nonmobile indoor machine.

Another alterative is under investigation at Stanford University, where the development of a superconductor FEL could lead to very efficient and compact models. In a superconductor system, the magnets are cooled to such low temperatures that the electric currents travel with almost no loss of energy. It will certainly be several years before the FEL technology is mature enough to be used for active service on the battlefield, but if the problems of size and technology can be solved, the frequency-agile FEL will be a prime candidate for tactical HEL weapon applications.

In principle, an X-ray laser beam could destroy electrical circuitry, possibly trigger some types of munitions, set off a nuclear bomb or render it inoperable, and make humans sick or even kill them. The preferred energy source for a very high power X-ray laser is a small nuclear explosion. This makes it almost impossible to contemplate a battlefield HEL X-ray laser weapon. Some research has been done by the Livermore Laboratories in the United States with optical laser-driven X-ray lasers. So far, the output power is modest compared to the input power. Thus, with the present technology, X-ray lasers are not candidates for battlefield HEL weapons.

Two excimer laser systems may be considered HEL weapon candidates—the krypton fluoride laser (KrF) emitting at 249 nanometers and the xenon fluoride laser (XeF) at 350 nanometers. The interest in using excimer lasers for weapons in a manner similar to the FELs has emerged out of the SDI program. Initially, the excimer work concentrated on the use of an HEL weapon mounted on a satellite to be used against nuclear ballistic missiles and warheads in outer space. Later stages of the program have placed the laser in a ground or underground station and reflected the laser beam by an orbiting mirror to the target in much the same way as with the FEL. While the FEL has the possibility of selecting an optimal wavelength, excimer lasers operate at only a few well-defined wavelengths. The basic problem is still to overcome atmospheric absorption and scattering. As the atmospheric effects are more severe at shorter wavelengths, the XeF laser at 350 nanometers should be a better choice than the KrF laser operating at 249 nanometers. A high-energy, Raman-shifted excimer laser at 353 nanometers was fired into space in March 1988 with a reported pulse energy of 400 joules, a duration of 0.5 seconds, and a beam width of 20 centimeters. This is believed to be the highest power laser pulse ever sent into space. Other recent experiments at the Los Alamos National Laboratory within the experimental AURORA program, which uses a KrF laser, show that some progress may be possible. The 249-nanometer AURORA laser delivered 1,300 joules to a 500-nanometer spot in pulses lasting 3 to 0.007 microseconds, corresponding to a total peak power on target of 10^{14} watts. However, this may be compared to the experimental solid-state NOVA Nd:glass laser at the Lawrence Livermore National Laboratory, which, during 1989, delivered pulses of 125,000 joules at 1050 nanometers and 10,000–20,000 joules in the third harmonic at 350 nanometers. Experiments with the NOVA at 350 nanometers are planned for the 70,000-joule region. The KrF excimer laser cannot presently compete with the solid-state NOVA Nd:glass laser, as the short wavelength of the KrF laser makes penetration of the atmosphere difficult, and this problem remains unsolved. Although the excimer lasers are the most powerful types in the ultraviolet spectral region, the problems with the very short pulses, the short wavelengths, and the special optics required for UV operation make the increase of output power to the same levels possible with the infrared chemical lasers a very difficult task. The highest average powers from excimer lasers are still much lower than can be obtained from infrared chemical lasers.

If we compare all of the alternative laser types for NEL weapon applications, a few remain as feasible short-term possibilities, but it is still doubtful if any cost-effective HEL weapons can be realistically fielded within the next 10 or even 15 years. If any idea of a battlefield HEL weapon still seems valid to staff planners, it will certainly be one that is based on the gas dynamic CO_2 laser, the electrically pumped CO or CO_2 laser, or the DF laser.

Both the iodine:oxygen and excimer lasers must be considered dark horses. It is questionable if any military requirement now includes plans for the destruction of hard targets. It may be that the really cost-effective solution for HEL weapons on the battlefield is to concentrate the R&D work on the more realistic and limited requirements of attacking sensors, many of which are extremely vulnerable to laser energy.

PREVIOUS HEL WEAPON PROJECTS

There are public reports that a target drone was shot down in experiments by the U.S. Air Force as early as 1969 using a primitive gas dynamic CO_2 laser. What has been more widely reported, and even shown on a film in public in 1982 at the annual Conference on Lasers and Electro-Optics (CLEO), is the shooting down of small, winged, propelled target drones as part of some 1973 vintage experiments conducted by laser scientists from the Air Force Weapons Laboratory at the Kirtland Air Force Base in New Mexico. They used a gas CO_2 laser of a few hundred kilowatts. The target drones were destroyed by detonating their fuel tanks and by cutting control wires. These experiments were certainly made under almost ideal conditions and only served the purpose of getting a basic knowledge of what could be done with an HEL weapon and what problems were involved. Detailed data and conclusions are still a well-kept secret, but it may be surmised in the end that these experiments simply proved that, in principle, laser weapons could work.

One of the first efforts to develop a prototype laser weapon was the Mobile Test Unit (MTU) by the U.S. Army in the mid-1970s. A 30-kilowatt electrically excited CO_2 laser was literally squeezed into a Marine Corps LVTP-7 tracked landing vehicle. In 1975, at Redstone Arsenal in Alabama, the MTU destroyed U.S. winged target drones as well as helicopter target drones. No real data are available to the public, but the experiments came to an end rather soon and have been reported as inconclusive. In the late

1970s, a German company, Diehl, worked on a concept for a laser weapon carried by a 28-ton armored tracked vehicle. It was based on a self-contained electrically excited CO_2 laser and may very well have been something similar to the weapon employed in the U.S. project MTU. The MTU was followed by the Close-Combat Laser Weapon (C-CLAW), dubbed ROADRUNNER by the U.S. Army. This was designed to attack enemy sensors, night vision equipment, and helicopter cockpits with a combination of rather low-powered Nd:YAG and CO_2 lasers. The restricted energy level and the military requirement to support combat units on the battlefield by attacking sensors both place this project in the category of low-energy laser (LEL) weapons, which will be described in more detail in the next chapter.

In 1978, the U.S. Navy conducted a series of tests as part of the Unified Navy Field Test Program at San Juan Capistrano in California, in which a chemical DF laser in the 400-kilowatt range destroyed some TOW wire-guided antitank missiles in flight. To direct the laser to this target, which was comparatively small and fast, a Hughes aircraft aiming and tracking system was used. In 1980, a captive UH-1 helicopter was destroyed by this laser system.

The U.S. Air Force placed a gas dynamic CO_2 laser in a Boeing NKC-135 cargo aircraft, dubbed the Airborne Laser Laboratory, and in 1981 tried to shoot down air-to-air AIM-9L Sidewinder missiles while airborne. These tests, performed at the Naval Weapons Center in China Lake, California, were a failure, and, as the planning had been made public in advance, the media could criticize the failure openly. The testing continued without any more media coverage, and finally, in May 1983, the 400-kilowatt laser shot down a number of Sidewinder missiles. The program was terminated in 1984, and the Airborne Laser Laboratory ended up in a museum. The aim, to prove that air-to-air and ground-to-air missiles can be destroyed in flight by airborne HEL weapons, had been validated, at least in principle. However, it must be remembered that this laser weapon completely filled a four-engine cargo airplane, and the experiment did not seem to offer any possibility of a weapon that could be carried as add-on equipment on a relatively small fighter to protect it from missiles. In any case, the results must have provided some clues as to how the problems of tracking a target and aiming the laser could be solved.

In 1981, the U.S. Army designed a Mobile Army Demonstrator (MAD), which was based on a small, compact DF laser. The demonstrator was used as a prototype for an air defense weapon against missiles which started at 100 kilowatts but was to be scaled up to 1.4 megawatts. The use of a DF laser poses some difficult problems. The exhaust gases are very poisonous and cannot be vented in the vicinity of friendly forces. The designers tried to solve this by using a closed system which collected the waste gases in a special tank. The tests ran until the project was omitted from the SDI program in the 1983–84 budget. However, development of the laser itself, renamed the Multi-Purpose Chemical Laser (MPCL), continued with U.S. Army funding of Bell Aerospace Textron.

One very interesting HEL development which has been the cause for much debate in the U.S. Congress is the Mid Infra-Red Advanced Chemical Laser (MIRACL) coupled with the Sea Lite Beam Director (SLBD). MIRACL is a DF laser with a 2.2-megawatt output at 3,800 nanometers. Sea Lite, later called Sky Lite, is the beam steering device for the laser. In the 1988 Strategic Defense Initiative Organization (SDIO) report to Congress, the MIRACL/Sky Lite was described as "the highest power HEL system in the free world." Some rather speculative thought-provoking demonstrations have been made with this system at the High Energy Laser Test Facility at White Sands Missile Range, New Mexico. On September 18, 1987, several vital components were destroyed on a Northrop BQM-74 airborne target drone, which then crashed. The laser test crew had to find, lock on to, and shoot down the drone, which was flying at a speed of 500 knots at an altitude of 1,500 feet. According to the press report, the system downed a Teledyne Ryan Aeronautical Firebee BQM-34S target drone at twice the range in November 1987. Two years later, a Vandal supersonic missile simulating a sea-launched cruise missile was forced down while flying at low altitude and at a range that was "representative of a real tactical scenario." According to the U.S. Navy, the test demonstrated that "HELs can be a real option for tactical warfare missions." However, the laser is presently too large for a ship, and, therefore, a smaller prototype system may follow, aimed at giving the Navy a shipboard laser weapon that would be able to destroy numerous antiship missiles at operationally effective ranges. Such a prototype is necessary to find out if it is really possible to solve the atmospheric problems in the humid environment at sea. When SDIO reported on the MIRACL/Sky Lite program in 1988, the objectives were given as the

> development and demonstration of a high-power local loop adaptive optics system for improvement of the beam quality of a multi-line infrared high energy laser; development and demonstration of a high power target loop adaptive optics system for ground to space atmospheric compensation in the presence of turbulence and strong thermal blooming; and performance of atmospheric propagation experiments to explore the conditions under which stable correction can be achieved and the degree of correction possible.

In other words, its purpose is to show that a really powerful infrared laser can be made to work as a weapon under more or less real battle conditions.

The future funding of the MIRACL/Sky Lite Program was heavily debated in the United States because of concern over the size (and cost) of the laser so that continuation of the program seemed in doubt for a while. Some statements made during the debate may be of interest. At one stage, when deletion of the beam director was suggested, the SDIO declared that it needed the MIRACL for its own missile vulnerability tests, similar to the test in which a laser beam destroyed the second stage of a pressurized Titan I rocket in 1985. Even if the SDIO may have had little real use for the MIRACL/Sky Lite as it wanted to explore the FEL, the U.S. Army had a growing interest in MIRACL for use in short-range missile defense experiments. Some military people urged the continuation of the program with three aims in mind: continuation of Navy anti-cruise missile tests, continuation of experiments on satellite vulnerability, and the tests and experiments cited by the SDIO. Finally, the project got funded for 1989.

Officials from TRW Inc.'s Space and Technology Group at Redondo Beach, California, have been urging the U.S. Navy ever since to fund a shipboard test program as the next step in the evolution of the HEL technology used in the MIRACL/Sky Lite program. The Navy presently has no funds for such a project, but, according to the Navy's Space and Naval Warfare Command, lasers could play a significant role in naval warfare in the future. As missiles approach supersonic speeds and incorporate stealth-like capabilities, the Navy will need the near-instantaneous targeting and killing abilities inherent in laser technology. According to TRW, their program is ready to move to tests at sea, because the laser already has been tested extensively and modified through development under the SDI program. The problem is the cost, amounting to several hundred million dollars, to deploy the laser system aboard a Navy test ship. Whatever the future of the MIRACL/Sky Lite program, it has certainly managed to create a heated debate in the United States over the viability of HEL weapon systems.

Research and development of high-energy laser weapon systems is proceeding also in France. The system named LATEX (Laser Associé à une Tourelle Expérimentale) consists of a laser in the 10-megawatt range coupled to an advanced aiming system commercially developed by Laserdot. The program was started by the General Delegation for Armament in 1986 and has advanced to a preliminary test carried out at Marcoussis in France over a range of 200 yards against a missile head and an aircraft fuselage panel. It has been reported that trials will now proceed in Landes, in southwest France, against a target flying at 300 yards per second at a range of $1\frac{1}{4}$ miles. LATEX may be similar in concept to the German air defense laser, HELEX, and the French Ministry of Defense has indicated an interest in cooperation with Germany. This kind of cooperation on other systems has probably been going on between these two countries for some time.

GERMANY'S AIR DEFENSE LASER (HELEX)

One of the most interesting HEL weapon projects is the German air defense system called HELEX, which is an industrial joint project between Diehl, Gmb., in Nuremberg and MBB in Munich. HELEX stands for High Energy Laser Experimental. The project is still in its early stages, although the initial work started in the late 1970s. MBB together with Diehl have been commissioned by the Federal Ministry of Defense in Germany to implement and study this experimental system as a continuation of the work done previously. In the following discussion, the term HELEX refers to the industrial conception of the final weapon to be delivered to combat units if the experiments are successful. The project is interesting, not only because a comparatively large amount of information has been made public so far, but also because it tries to meet a precise military requirement. Since this is not only a research program but also a very extensive development program aimed at producing a well-defined laser weapon for a future battlefield, it will be described in detail. The idealized conceptualization is given in Fig. 5.1.

Germany has a long common border with Poland and Czechoslovakia, which were Warsaw Pact (WP) countries, and the distances from important targets inside Germany to WP air bases and missile sites were very short. The time between an airborne attack launched from the WP air bases across the border could be extremely short, lasting only minutes. Thus, Germany was very vulnerable to low-level air attacks by combat aircraft missiles and standby weapons with the capability of engaging targets automatically. However, the distances are still comparatively short, and, even though the warning time is slightly longer, this limited

----

FIGURE 5.1. High-energy CO_2 laser system. The laser energy is directed toward the target by a highly controllable large mirror, which, on its scaffolding, can go over buildings, trees, and other ground obstructions. Photograph courtesy of MBB/Diehl.

distance will still be a problem for Germany's air defense. The present-day German air defense is heavily dependent on ground-to-air missiles, fighter planes, and sophisticated chains of radar stations which feed the command and control system with information. In spite of all the money spent so far on this very complicated air defense system, it may be insufficient to counter future threatening situations in which the other side will use an increasing amount of more and more sophisticated electronic countermeasures. Air defense laser weapons could be one way to achieve the extremely short warning and engagement times that Germany will eventually require.

The main component of the HELEX is a gas dynamic carbon dioxide laser which emits an average beam power of several megawatts over the specified mission time. To carry the laser and all of its accessories, the basic chassis from a German tank, Leopard 2, has been suggested. The supply tanks for gas, water, etc. are used for the laser fuel, while the laser itself and its coolant water are carried in the chassis. As laser weapons have a direct-line-of-sight action, it is important to position the laser beam above the tops of surrounding trees and buildings. This problem is solved by using an elevator platform to carry a focusing mirror of more than one yard in diameter along with the passive surveillance and target acquisition system. The area of coverage of the HELEX will also be greatly increased by the elevated platform, since the time between the identification of a target and the laser hit is very short, and it may be possible to engage very low flying targets that quickly appear and disappear out of the immediate field of view.

A relatively simple technical principle has been used for the HELEX. The high-energy gas dynamic laser employed does not need a heavy and complicated gas pump or flow system nor does it require sophisticated cooling. The fuel is a common hydrocarbon burned together with a nitrogen compound oxidizer, both of which can be easily carried in the liquid storage containers. The hot gas flows at supersonic velocity through a comb of very fine nozzles, expands, and is transformed into the population inversion state required to amplify the laser energy. The gas then flows at supersonic speeds through an optical resonator (mirrored cavity), where stimulated emission occurs, and the laser beam is finally created. The beam leaves transverse to the gas flow direction. The used, nontoxic gas is vented into the atmosphere through a diffuser. At the same time, the exhaust gas carries off most of the waste heat. Overall, the function of the laser is similar to that of a rocket engine.

The emitted power of the high-energy gas dynamic laser is proportional to the amount of fuel used. The research to date indicates that the dimensions of even very high energy laser equipment will remain within acceptable limits from a technical point of view. The fuel consumption per laser shot corresponds roughly to the weight of a guided missile, but the fuel consumption of future-generation systems should be lower. If these estimates are correct, an HEL weapon like the HELEX should be able to fire something like 50 laser shots with the amount of fuel (5–10 tons) carried in the tank.

The wavelength of the HELEX system may be either 9,350 or 10,600 nanometers. Most reports on the system indicate a wavelength of 10,600 nanometers. However, the shorter wavelength may be a more appropriate choice, since the larger the focusing mirror is relative to the wavelength, the smaller the focal spot and the higher the energy density will be. Obviously, the desired effect requires as high an energy density as possible.

The optics of the HELEX must cope with the difficult task of focusing enough laser energy on the target to destroy it in the air or cause it to crash. This has to be done on the battlefield even when the atmospheric conditions are unfavorable and at a combat range of at least five to ten kilometers if the HELEX is to be cost-effective within the air defense concept.

Only mirrors suitable for use at the wavelength and high power levels of this system can be used to direct and focus the beam. The use of transmission optics such as lenses is not very feasible due to their high cost and fragility, and, in any case, the HELEX will probably damage any lenses to some extent. The reflector at the top of the elevated platform is a concave mirror with a diameter of more than one meter. To achieve a sufficient effect at the target range, compensations for atmospheric turbulence, blooming, and other disturbances to the laser beam inside and outside the system are planned with an adaptive mirror. The mirror surface can assume the required shape and the correct axial angles with the aid of numerous piezoelectric (small electronic) actuators exerting mechanical forces on the mirror back. To enable the mirrors to withstand the HEL beam, a cooling liquid flows through fine channels on the rear of the mirror. Compensation by adaptive optics may double the range possible with a rigid mirror system.

The information necessary to control the mirror surface shape is furnished by the laser beam reflected by the target; thus, the beam itself becomes a sensor element in the closed control loop by which the target is tracked. It is a difficult problem to achieve a really high precision laser beam, and it is necessary to keep a focused beam on a single location of an extensive target for a considerable time. If a target moves at the speed of sound (Mach 1) and the beam must be coupled with it for, at least, a half a second, during this time the target will move nearly 105 yards. Keeping the laser on the same spot may be done either by using the variable reflection characteristics of the target or by a procedure where the deviation of the beam center from reference marks on the target is used as the control signal. Diehl has demonstrated this procedure by means of a rotating aircraft model.

It is not only necessary to keep the beam directed to the same spot on the target, but it is also a prerequisite for the HEL system that the beam can be focused correctly. Basically, the mirror at the top of the elevated platform functions just like a burning glass which concentrates the energy of the sun to such an extent that combustible material catches fire. The advantage of the coherent laser energy is that it can be focused sufficiently over distances of many kilometers to produce thermal effects at the final site. Adaptive optics can be used to focus the beam continuously, even as the target changes its position.

The HELEX will have some type of a passive surveillance and target acquisition system, such as satellite monitoring, which will probably cover the entire hemispherical air space of the protected zone and permit tracking of numerous targets simultaneously. This is also the prerequisite for sequential engagement of targets by the laser weapon without any delay. The passive target acquisition makes radar surveillance and tracking unnecessary, and, as a passive surveillance system is used, it may be very difficult for an airborne attacker to find and counter the system beforehand by any electronic countermeasure activity. The HELEX will make it possible to carry out identification, threat analysis, and target selection and finally to hold the beam on the target on automatic, or, if desired, part of the sequence can involve a human operator to select target priority. However, the choice of the best or, at least, a suitable spot to hit on the target has to be done automatically to cope with the time constraints.

If the research and development of the HELEX air defense laser weapon is successful, battlefield commanders will have a powerful tool to cope with highly threatening situations. One air defense HELEX could effectively control an area against multiple low-level, high-speed attackers with comparatively low operating costs. The effective range will be dependent on atmospheric conditions. Under very favorable conditions, the range against aircraft, helicopters, and missiles would be up to 6 miles; this would be reduced to 3 to 4 miles in the normally heavily polluted atmosphere over a battlefield. Due to the extremely short time for target detection, tracking, slaving, and firing, it would be possible to engage many targets in rapid succession. If one HEL weapon is defending a facility that is attacked by a squadron-sized enemy force, the laser weapon may very well shoot down all aircraft during their first attack. Reloading is simple; there is no minimum range, and different types of targets do not require the use of different types of ammunition. The main limitation of such a weapon as the HELEX is the reduction in range of the system under very poor weather conditions or when the pollution on the battlefield is extremely heavy. It is difficult to quantify these limitations, but it is obvious that the HELEX will not replace conventional gun and missile systems, not even at distances well within its range. Such HEL weapon systems will only be able to complement existing air defense systems. However, the survivability on the battlefield of a HELEX type system compared to a system dependent on radar technology will be very high, since the passive localizer will not reveal itself. Also, the mobility of a 20–40-ton tracked HELEX system will be high, and it will be possible after terminating one firing action sequence to change the location of the weapon quickly.

Many problems still must be solved before it is even possible to decide if the HELEX concept is a valid one. To date, tests have only been done in the laboratory. The scaled-down experimental weapon paid for by the German Ministry of Defense will not be available until 1993 or 1994. If this weapon is a success, and if it is possible to solve all of the very difficult problems, the development of a final air defense high-energy laser weapon based on the HELEX concept may start in the mid-nineties and should be completed about ten years later. This means that theoretically such a weapon could be produced and handed over to the combat units at the beginning of 2005. Due to the technological difficulties involved in this concept, even such a distant delivery date may be overly optimistic.

Other countries have begun developmental work on possible laser weapons along similar lines. In France, several companies together with the French National Aerospace Research Agency (ONERA) are working on a HELEX-like experimental HEL weapon. There have also been some reports on a possible collaboration between France and Germany. In the United States, a similar idea is currently under investigation in the JAGUAR project.

The military specifications for the HELEX weapon are really very ambitious, and this, along with the technological difficulties, is the main reason for the high costs and the very long time necessary for research and development. It is debatable whether or not it would be more cost-effective to limit the requirements to simply damaging some very sensitive parts of the target such as sensors, canopies, and radomes and leave the actual destruction of the platform itself to conventional antiaircraft guns and missiles. This would mean that, up to 6 miles, much less energy would be required, and the sensitivity to the atmospheric conditions should be less. Such a weapon could possibly be fielded much earlier and at a significantly lower cost. Of course, there are even some limitations to this less demanding military requirement. Some of the targets are not all that dependent on their sensors, and, even if they are, it may be possible in the future to make the most crucial sensors insensitive to the effects of laser energy. Whatever the future holds for the HELEX, the fact remains that a high-energy air defense laser weapon capable of outright destruction will be expensive to develop and manufacture, and it will take many years before such a weapon can be successfully fielded. It is very possible that the whole idea will be abandoned because it simply proves technically impossible or just too expensive to implement.

HEL WEAPONS IN THE SOVIET UNION AND THE NASCENT REPUBLICS

Very few facts are known to the public about the research and development of HEL weapons in the Soviet Union or its surviving constituents. Some official reports and statements are available as well as some material by independent writers, but most are of a very general nature. This, of course, is not surprising; all work on laser weapons in the West is shrouded in security, and very few facts are made public. This is even truer in the Soviet Union.

However, the fact that so many papers on high-energy lasers and their effects have been published in Soviet scientific journals is an indication of the amount of work done in this field and, thus, reveals the strong interest of the Soviet Union in this technology. The papers, of course, deal only with basic laser technology and not with the details of developing laser weapons. There have been unconfirmed reports of the installation of a high-energy chemical laser on a Kirov-class cruiser. The HEL weapon was said to be successfully used against the sensors of sea-skimming missiles out to a range of 10 miles. If this is true, it may have been some kind of experimental installation, as the existence of such a weapon has not yet been confirmed. There have also been descriptions of the Soviet research facility at Sary Shagan in Kazakhstan following a visit by a delegation of U.S. scientists in 1989. A beam director with a diameter of approximately 1 yard was connected to a ruby laser and to a carbon dioxide laser and, according to U.S. analysts, had been used in tests against both aircraft and satellite targets.

The most powerful laser at Sary Shagan was reported to be a 20-kilowatt CO_2 laser. Another U.S. delegate from the U.S. House Armed Services Committee in 1989 reported the existence of a previously unknown high-energy megawatt-range Soviet laser, seen when he visited the Kurchalov Institute of Atomic Energy in Troisk, a center of scientific research south of Moscow. It was a 1-megawatt CO_2 laser, and the Soviet officials claimed that it was unique in the country and that they had been operating it for several years.

In January 1987, the Pentagon published an edition of *Soviet Military Power* including a photo which was identified as a laser device on a Soviet destroyer that has been used in the past against Western patrol aircraft (U.S. Department of Defense, 1987). If the picture really shows such a laser, it could only be a low-energy laser with no capability of destroying aircraft or missiles but rather with a blinding effect for sensors and eyes. The same publication states that

> the Soviets have built high energy laser devices up to the 10 megawatt level and generally place more emphasis on weapon application of lasers than does the West. In doing so, the Soviets have concentrated on gas dynamic and electric discharge lasers. They have not attained a high power output for chemical lasers as the West.

There is no real proof or even any strong indication that the development of high-energy laser weapons is in a more advanced stage in the former Soviet Union than in the West. Scientists in the former Soviet Union are probably working hard along the same lines as Western scientists within the Soviet version of the SDI program and within the various concepts for tactical use of HEL weapons on the battlefield. Based on the work done so far in the West, it may be concluded that the fielding of HEL weapons is as many years away for the former Soviets as for the West. However, as we will see, the situation may be different when it comes to low-energy laser weapons.

ASSESSMENT OF LASERS FOR FUTURE HEL WEAPONS

The development of HEL weapons will certainly continue in both the East and the West based on the present as well as new concepts of HEL weapons as long as there seems to be a reasonable possibility of solving the problems involved in fielding a cost-effective system. Military staffs and research centers will probably stick to the concept of air defense of important targets and of attacks on sensor systems as the main area of use. As long as the Eastern bloc and the Western SDI programs allocate substantial financial resources to laser technology, there will continue to be spinoffs to tactical HEL weapon projects for the conventional battlefield.

Besides the CO_2, CO, and chemical DF HEL weapons exemplified by the German HELEX and the U.S. MIRACL systems, research and development of HEL weapons will also progress based on other laser concepts such as the U.S. ground-based free-electron laser (GBFEL). The FEL may be a future choice in HEL weapon applications for antisensor and air defense tasks. According to some reports, very efficient FELs are on the horizon, and, with the new superconductor technology, very compact and efficient FELs may soon be possible. Another report gives the impression that low-power tunable FELs are under development as short-range high-energy weapons. Some work will certainly proceed, although the statement that "in the weapons field, contemporary

FELs promise to become the germ of the ray gun of the future which hurls powerful bolts of energy at the enemy" seems a bit premature.

Another laser test program of the U.S. Air Force involves a moderate-power Raman-shifted excimer laser device (EMRLD). The work is performed at the Kirtland Air Base Weapons Laboratory in New Mexico, and the stated goal is to produce more than 5,000 watts at 100 pulses per second. Some of the technology developed in this way may finally be used in real battlefield systems.

A development project is also under way, or, at least, planned, with an iodine:oxygen laser. This is indicated by the worries of some administrators that MIRACL would compete for funding with other beam projects such as iodine:oxygen, chemical, and excimer lasers.

Thus, although there are some promising lasers that may form the basis for HEL weapon systems usable on the conventional battlefield, there are still some unsolved problems. Even if the laser can achieve a sufficient energy output, the atmospheric conditions still severely limit the practical use of HEL weapons. So far, no HEL weapon program seems to have solved this problem, and it is still somewhat uncertain whether or not it is really possible to do so. Techniques such as the use of an active mirror that can adapt instantly to varying conditions have still only been demonstrated in small-scale models.

CONCLUSIONS

The interest in HEL weapons is high in many countries due to the great advantages that they theoretically offer. The HEL energy travels at the speed of light with a flat trajectory and acts almost instantly on the target. These are qualities which, if they can be used in an air defense situation, for example, can neutralize even the smallest and fastest missiles. Therefore, it is no wonder that HEL weapons continue to fascinate military staffs and military laboratories sufficiently to fund further research and development. Also, in some countries, spinoffs from SDI programs have contributed heavily to knowledge about and interest in HEL weapons.

However, the use of HEL weapons within the atmosphere presents severe problems. Thermal blooming, turbulence, scattering, and absorption all have very negative effects on the laser beam, and these difficulties grow rapidly with increasing range. Battlefield conditions mostly imply a heavily polluted environment, which adds still more problems. We conclude that, even if adaptive optics and other technological problems can be solved in the laboratory, still no operational weapon is within sight. To date, it is not even known if a usable, cost-effective solution is possible in the distant future. Assuming that all the problems could be solved satisfactorily in the near future, the earliest HEL weapons could not be handed over to combat units until the beginning of the next century.

The difficulties associated with the development of HEL weapons are, to a great extent, generated by the very tough military specifications to literally burn holes in airborne targets at long ranges. This requires megawatt lasers with a range of at least 3 to 6 miles in a hazy atmosphere. If the requirements could be limited to attacking sensors and other highly laser-energy-sensitive parts of the target instead, it would be possible to field an HEL weapon earlier and at much lower cost. Since most important airborne targets are highly dependent on electro-optical devices and other items sensitive to laser damage, a medium-sized HEL weapon could very well turn out to be a cost-effective weapon.

SIX

Low-Energy Antipersonnel and Antisensor Laser (LEL) Weapons

BACKGROUND

Laser devices have found extensive military use in the last 20 years. A steadily increasing number of laser designators, range finders, and other systems are presently being fielded. Today, such low-energy laser (LEL) systems are common battlefield tools in most combat units. Some of the lasers in their present configurations may already be technically suitable for use against the eyes of enemy soldiers and even against some of the enemy's electro-optical sensors. This is especially the case with target designators and most range finders. However, it is questionable if it is worthwhile to use present-day lasers in any systematic fashion as effective weapons. Even if range finders and designators are dangerous to eyes and sensors, most are not designed to be powerful enough to be considered cost-effective laser weapons. In spite of this drawback, today's lasers certainly represent a threat to soldiers using magnifying optics, and the deliberate use of present-day lasers as weapons will create, if not a serious threat, at least, a nuisance hazard in that they will deter forward observers and others from using magnifying optics. The future potential of present laser devices as weapons is clearly revealed by the existence of the very far-reaching safety regulations we have established already. The nominal ocular hazard distances (NOHD) for some typical laser devices listed in Table 6.1 illustrate the magnitude of the problem.

Laser safety regulations are about the same all over the world, and most originate from research done in the United States, starting from when the first laser was invented. The hazard distances are usually calculated based on the worst-case scenario—an eye that takes in as much light as possible and a laser beam that is focused on the most sensitive part of the eye. A safety factor generally around 10 is then included in these hazard calculations. However, even with these distances inflated by the safety margin, the hazard is still sufficient to be a considerable threat to the eyes of soldiers on the battlefield and certainly even more so if magnifying optics are used. Even if the eye damage causing a moderate loss of vision or actual blindness is mostly an incidental effect of the use of range finders and designators, their devastating results may tempt commanders and soldiers to deliberately use their tracking devices as offensive weapons. Against this background, it is not surprising that, according to experts, present laser devices are deemed likely to be used against eyes and sensors for whatever benefit they can yield.

TABLE 6.1.   Nominal Ocular Hazard Distance (NODH)
for Some Typical Lasers

	NOHD in miles		
	Naked eye	8× Optics	13× Optics
Tank range finder, ruby laser	10	23	80
Tank range finder, Nd:YAG laser	4	16	22
Portable range finder, Nd:YAG laser	1	6	9
Portable target designator, Nd:YAG laser	8	25	33
Airborne target designator, Nd:YAG laser	25	46	80

Laser accidents resulting in injuries to personnel have occurred regularly since the invention of the laser. Some of these accidents are not connected with the laser beam itself but have been caused by related hazards such as the high voltages used to power the laser. Most accidents involving laser radiation have occurred in laboratories. Reports of outdoor accidents with military equipment have so far been rather rare. This will probably change with the widespread deployment of hazardous laser devices in combat units.

We can gain some knowledge about the possible effects of laser injuries in combat from laser injuries that have occurred accidentally. Laser accidents have been reported from several countries; the total number is probably several hundred or so. However, only a few have been described in sufficient detail to allow for careful analysis. Most accident reports have been presented at various conferences on laser hazards. Six accidents occurring in the United States were reported at the International Conference on the Biological Effects of Large Dose Ionizing and Non-Ionizing Radiation held in China in 1988, with the suggestion that the information learned from these accidental laser exposures would prove to be a valuable adjunct to experimental animal research programs on laser bioeffects. The accidents revealed short-lived transitory effects at exposure levels well below the maximum permissible limits. Such data are not normally obtained from animal studies. Psychological reactions to laser injuries may also be studied in human accidents quite easily as opposed to the difficulty of monitoring such effects in animal experiments. It is clear that accidental injuries, when properly reported, can serve as a tool for predicting psychological reactions of great interest to modern military establishments.

There have been some reports, and even more rumors, about the use of laser weapons or other laser devices against soldiers in recent conflicts. It has been unofficially reported that Soviet chemical lasers were used in the late 1970s to blind Chinese soldiers during the conflict between China and Vietnam. These rumors had already been referred to some years ago at a U.S. Army conference on combat ocular problems held in San Francisco. One of the participants commented on a *Newsweek* magazine report stating that carbon dioxide lasers had been used against soldiers during that conflict and that many burn casualties were found in several Cantonese hospitals. It was also mentioned at the conference that there were several cases of eye injury, but no documentation was forthcoming. The recent Soviet conflict in Afghanistan gave rise to a press report that "the incidence of laser blinding was becoming so frequent that the area has become a concentrated test-bed." The same report stated that an early U.S. laser weapon program evolved from an initiative to give U.S. pilots some way to knock out the optics of Soviet-made North Vietnamese antiaircraft systems. This system worked when initially deployed, but the North Vietnamese shot down one of the laser-equipped airplanes and presumably turned the system over to the Soviets to work out a way to nullify it. According to David Isby, author of *Weapons and Tactics of the Soviet Army*, laser range finders used by Soviet combat units in Afghanistan blinded at least one guerrilla soldier who was using binoculars. Isby also states that this soldier received treatment in the United States. Reports have also been received of lasers being used against human eyes in the Iran–Iraq war during the mid-1980s. The Iraqis could only have employed Soviet laser range finders or other laser devices for this purpose. Although rumors about the use of lasers as weapons are a frequent occurrence, actual reports of laser attacks are sparse and mostly inconclusive. Some of them may originate from a soldier who has mistaken an ordinary flashlight or searchlight for a combat laser. Other reports may come from soldiers who were unintentionally exposed to the beam from a laser range finder or a designator. Some, but probably very few, reported laser attacks may be instances of the deliberate use of laser range finders or designators as anti-eye or antisensor weapons. Reports from recent military conflicts contain no proof whatsoever that low-energy laser (LEL) weapons have been fielded and deliberately used.

Some peacetime laser incidents involving combat units from different nations have been reported over the years. One of the most publicly discussed is the Pentagon report from 1987 stating that a Soviet ship used a laser against two U.S. military aircraft flying over the Pacific Ocean in an area north of Hawaii. The U.S. airplanes, one a Navy P-3C Orion antisubmarine aircraft and the other an Air Force WC-135 weather monitoring aircraft, were observing the intended target area for a test launch of a Soviet intercontinental ballistic missile. There was a report that a shipboard Soviet laser temporarily blinded the woman copilot of the WC-135 without any damage resulting to either of the aircraft. The Defense Intelligence Agency claimed that the copilot was impaired for a period of 10 minutes. She was certainly examined very carefully afterwards for any signs of eye damage, but only "inconsequential" effects were found. The detailed information on this incident has been classified, which makes it almost impossible to second-guess what really happened. If the incident occurred in darkness or twilight, a low-energy laser may very well have been used. However, if the eyes of the crew were not dark adapted, it is unlikely that the light coming from the ship originated from a laser weapon. It is practically impossible to flash blind a person in broad daylight without also causing some lasting damage to the eyes. Flash blinding without any damage is only possible when the eye is dark adapted and, thus, much more sensitive to the incoming laser light. It is also very surprising that no one in the crew other than the copilot was affected by the laser or reported a bright light on the ship by indirect viewing. If the light was strong enough to flash blind one person in the crew, it is most unlikely that no one else noticed it or was affected by it as well. There must be more conclusive information in the classified part of the report, because the American authorities have publicly taken the position that it was a laser. Furthermore, they have stated that there have been other instances of pilots being blinded, although none permanently as yet, from very "powerful Soviet laser systems aboard vessels."

A treaty not to use laser weapons in peacetime against each other was signed by the United States and the Soviet Union in 1989. This treaty seems to be an acknowledgment that both countries already had at that time laser devices that might be deemed dangerous to both material objects and human beings. This treaty is discussed in more detail in Chapter 8. During the negotiations of this agreement, an anonymous U.S. official concluded that both countries would rather not admit that they were developing tactical laser weapons, so it would be safe to predict that the treaty would simply avoid the issue by not mentioning them at all. In spite of this treaty, four laser incidents were reported in October and November 1989 involving Soviet ships and U.S. aircraft. According to a U.S. Department of Defense official, one incident described as a "possible laser illumination" may have resulted in "visual injury" to a U.S. crew member.

There can be no doubt that both research and development of LEL weapons are progressing rapidly mainly to fulfill the urgent military needs for countering the many different sensors on the modern battlefield. All indications show that progress in the 1990s will be rapid and dramatic and will ultimately change the tactics and doctrines on the battlefield sufficiently to require new methods of protection and combat training.

Some of the LEL weapons, and perhaps most of them, will undoubtedly have an antipersonnel effect by damaging and destroying the most common sensor on the battlefield—the human eye. The eye is not only the most common sensor but one of the most sensitive. The effect of the use of lasers as antipersonnel weapons is not limited only to the biological damage, but it may also have far-reaching psychological consequences for the people at risk.

It will certainly be tempting to the military to increase the energy of range finders and designators and use them deliberately and systematically as anti-eye laser weapons. The expected military advantages may even lead to the development and large-scale fielding of LEL weapons that are exclusively or mainly designed to harm eyes. Such a deployment and systematic use of anti-eye laser weapons will probably mean mass eye injuries, severe psychological effects, and a heavy burden on both wartime and peacetime medical resources.

The question as to whether anti-eye laser weapons should be

fielded is a very sensitive one. Negative reactions from the public in several nations indicate that such antipersonnel laser weapons may not be easily accepted. Another question to consider is whether antipersonnel laser weapons are in accord with international law. However, as long as the targets are nonhuman sensors or some other military equipment, there will certainly be no problem with public opinion or international law. Such LELs may even not be considered to be weapons but rather systems or devices used as electronic countermeasures. The question of the legality as well as the morality of anti-eye laser weapons is dependent on what the target appears to be. The target may be military equipment, yet may still be capable of damaging eyes. What would be done in this case?

LOW-ENERGY LASER WEAPON TARGETS

As mentioned before, it is only possible to use LEL weapons against targets that are highly sensitive to laser light. A prime target is the human eye, either naked or behind magnifying optics. When the vision of a soldier is affected by laser radiation on the battlefield, the effect varies depending on the soldier's distance to the laser and the laser beam's properties. The least destructive effect is discomfort glare. This means that the soldier cannot see at all or is very much disturbed as long as the laser is emitting visible light directed into his eyes. In many ways, it is the same effect that an automobile driver experiences at night when meeting another car head-on with high-beam headlights. The next step up on the scale of visual disablement is flash blindness, which means that the affected soldier, even after the exposure, cannot see for a shorter or longer period due to the afterimage and loss of sensitivity from the flash. Discomfort glare and flash blindness can occur without causing any permanent damage to the eyes as long as the soldier's eyes are dark adapted when exposed to the beam at night, and the energy level used to achieve flash blindness is very low. If, on the other hand, the energy level is sufficiently high, the soldier will be permanently damaged by retinal burns and intra-ocular bleeding. In the worst case, the resulting blinding will be permanent. If a military policy decision is made to design anti-eye LEL weapons, it is likely that the military requirement will be to cause as much damage as possible to impose a maximal burden on the enemy.

The modern battle cannot be fought without the assistance of an increasing number of electronic eyes. All high-technology weapon systems for detection, identification, tracking, and hitting various targets depend on different types of sensors. Night fighting is an almost impossible task without electronic devices that make battlefield surveillance and weapon engagements possible. Some sensors can see through fog, heavy rain, and battlefield smoke; other sensors can detect the heat from tank or aircraft engines. The sensors located within the intelligence control command systems are more and more a prerequisite for effective leadership in battle. Future high-technology systems such as armed helicopters and other aircraft, ships, armored fighting vehicles, and missiles will be even more dependent on sensors. The increasing use of sensors also means that there is a rapidly growing interest in developing and fielding electronic countermeasures that can blind or destroy the electronic eyes of the enemy. LEL weapons offer the very important advantage of possibly countering and destroying the optics and the electro-optical sensors on the future battlefield.

Heat-seeking missiles are dependent on infrared sensors for the detection of heat from the intended target. Such missiles are used in air-to-air combat, where the sensor guides the missile to the heated exhaust gases from a jet engine or to some other hot spot on the target. Similar missiles are used in ground-to-air applications within air defense systems. The shoulder-launched U.S. Stinger was used in Afghanistan and is a good example of such a heat-seeking missile. These sensors may be blinded, tricked, or destroyed by a laser beam locked onto the sensor before the missile reaches its target. Highly valued aircraft could, in the future, be protected by body- or wing-mounted pods containing antisensor weapons.

Thermal viewing systems are used by tanks and other armored combat vehicles, armed helicopters, forward artillery observers, and many other combat systems. They allow the gunner or operator to see through darkness, fog, smoke, screening camouflage, and thin vegetation such as bushes. The sensors detect differences in temperature, and a thermal picture or outline can be seen by the operator. It is possible to detect the lenses in thermal sights using a laser radar, which can register the reflection coming back from the thermal viewer. An LEL weapon may then be directed by the laser radar to fire automatically at the thermal sight and destroy it. Combat units equipped with such a weapon may easily be able to turn a night fighting engagement in their favor.

Another important and even more frequently used sensor, the image intensifier, is for night fighting and is extremely sensitive to visible light. The image intensifier is designed to intensify even very small amounts of visible light. Its picture is presented to the operator on a screen inside the intensifier, which, overall, may be the size of a large pair of binoculars. A laser beam within the visible part of the spectrum may easily destroy the sensitive detector at very long distances. Again, in this case, the possible countermeasure would be a combination of laser radar to detect the optics and an LEL weapon to damage the sensor behind the optics.

Some missiles have low-light television systems fitted to their front ends. A television picture of the target area is sent to the operator by an optical fiber or radio link, which then directs the missile. This sensor is extremely sensitive to laser light in the visible spectrum, and, thus, for a missile countermeasure, it would be sufficient to blind the sensor until the operator loses control of the missile. Low-light television systems are used for several different tasks on the battlefield, and most such sensors can easily be countered by an LEL weapon.

Some missiles use the beam riding technique, and these may also be countered by a defending LEL weapon. The hostile beam can be discovered by a laser detector which then traces the direction to the laser source. Another way to detect the hostile laser is with laser radar. In any case, once the position of the enemy laser site is discovered, it is possible to automatically fire a counter laser beam to destroy or disturb the enemy laser transmitter or some vital electro-optical equipment necessary for its proper operation.

At the upper end of the power range of LEL weapons is the capability of heating and distorting or cracking the glass lenses of optical systems. This effect is called crazing and is caused when the heat buildup and subsequent cooling in the glass surface creates uneven stresses in the glass sufficient to crack it. The result is a frosted effect, making it impossible to see through the glass lenses or vision blocks (glass windows) in tanks. Such targets may be affected at long ranges, and the optics can be crazed in less time than is needed to blink an eye.

It is obvious that optics and electro-optical sensors can be easily attacked by an LEL weapon. The laser radiation may craze the optics, or, if the wavelength is one which the sensor system is designed to accept, the laser energy will be transmitted through the system itself, destroying or jamming vulnerable detectors. Sensors are normally designed to handle very small amounts of radiation and cannot accept the high intensities of radiation achieved by even very low energy lasers. The sensor may be damaged up to a certain range or temporarily jammed at even greater ranges.

LEL weapons designed and used for antisensor purposes may be called optical and electro-optical countermeasure systems instead of laser weapons. However, low-energy lasers used against human beings are certainly weapons, and the possible use of an antisensor laser system against people will obviously place even this system in the weapons category.

LASERS FOR LOW-ENERGY WEAPONS

In contrast to the very limited number of systems in the HEL category, a variety of conceivable lasers for LEL weapons have been proposed for each military mission. They differ with regard to the choice of lasing material, pumping method, tracking method, and energy source. It is possible to find a suitable laser for nearly every LEL weapon application among the hundreds of different laser types available, even if most of the lasers are suitable for only a few specific LEL weapons.

Lasers for Anti-Eye Weapons

An anti-eye laser weapon that is not designed to cause only discomfort, glare, or flash blindness but rather to cause permanent damage leading to immediate incapacitation and severe permanent visual impairment or permanent blindness could either be optimized for attacking the outside of the eye, the cornea, or the inside of the eye, namely, the retina.

It is possible to severely damage the cornea with a laser that uses a wavelength which is not normally transmitted inside the eye. If sufficient average beam power is used, the cornea will suffer a damaging burn which looks like a round white spot in the pupil. This is a very painful injury which would lead to immediate incapacitation of the soldier and at the same time cause the cornea to become opaque and completely block vision. Such an injury may be only temporary, but it requires immediate medical attention. To inflict this type of injury, it is likely that the longer infrared wavelengths are the best ones. A medium-powered CO_2 laser operating at 10,600 nanometers or a chemical deuterium fluoride (DF) laser operating at 3,800 nanometers would be especially effective. However, the amount of energy required to cause burns to the cornea is rather high, 1 joule per square centimeter. If this amount of energy is to be delivered to the cornea in the short time before the eyelid shuts, it is necessary to have a laser with an average power of about 10 watts per square centimeter directed onto the cornea. Such lasers are outside the LEL category; they are more like flamethrowers, well up in the HEL range. If the goal is to use LEL weapons to damage vision, it is much easier to use that part of the spectrum which is transmitted by the optics of the eyes, as the energy levels required to enter the eye are then extremely low and fall well within the LEL range.

There are many low-energy lasers operating at one or more wavelengths within the retinal hazard region from 400 nanometers to 1,400 nanometers. Much work is presently in progress on designing lasers that are safe to the eye for military and public use. These lasers operate outside of the retinal hazard region. An example is the erbium:YAG laser at 1,540 nanometers. Also, there is a trend toward the development of military lasers for training, range finding, and target designating that are safe to the eye so that laser equipment can be used in two-sided military training exercises without any risk of eye injuries.

The effect of the laser energy within the retinal hazard region varies with the wavelength of the laser. The effect also depends upon the pulse energy, pulse length, pulse repetition frequency, and any individual variations in the retina. This variation can be seen when you compare the eye damage from lasers at 530 nanometers (frequency-doubled Nd:YAG), 690 nanometers (ruby), and 1,060 nanometers (Nd:YAG). If the viewer is using binoculars through a clear, nonturbulent atmosphere, bleeding inside the eye occurs for 530 nanometers out to 2,900 meters, a distance only 20% further than for 690 nanometers but 10 times as far as for 1,060 nanometers. This is mainly due to the differences in absorption at different wavelengths by the sensitive structures in the eye.

It is obvious that the LEL weapon designer must select a wavelength and other laser properties carefully in order to obtain the combination that will make a particular weapon as effective as possible. It is also necessary to take into account any possible measures that might have been taken for the protection of the target. For example, it is possible to have a filter which gives full protection against a ruby laser. This, of course, means that it is not very cost-effective to design a low-energy ruby laser weapon when this filter is cheap and easily available to most soldiers. The weapon designer will have to counter this possible protection either by using a tunable laser, termed an "agile" laser, or by using one or more wavelengths which are in a part of the spectrum for which a filter is very difficult to build or must be so dense in order to provide proper protection that it will be difficult for the soldier to see through it sufficiently to fight well.

It is possible to point to some specific lasers that could be used as an anti-eye laser weapon alone or in combination with one or more other lasers. The Nd:YAG laser with a wavelength of 1,060 nanometers and the frequency-doubled Nd:YAG laser with a wavelength of 530 nanometers are both interesting candidates, although it is possible to protect the soldier using filters which block each of these wavelengths. Protection filters have also been used with the ruby laser at 690 nanometers, but it is much less feasible to protect against all three of these wavelengths using a single filter, thus making "protected" targets vulnerable to this combination of weapons. However, it is inefficient to use three lasers against one target instead of finding one that is strong enough to do the job alone or that cannot be protected against with a filter.

If the designer wants to use one or more lasers in the visible part of the spectrum, where it is very hard to give protection, yet still allow for good vision by his own troops in a close combat situation, argon lasers with wavelengths at 488 and/or 514 nanometers and titanium–sapphire lasers tunable between 660 and 980 nanometers are good possibilities. It is possible to cover an even larger part of the spectrum by adding the effect of frequency doubling and Raman shifting.

One of the alternatives for a tunable laser is the alexandrite laser, which can be continuously tuned from 700 to 815 nanometers with the maximum energy achieved at 755 nanometers. Such a laser, manufactured by Allied-Signal's Electro-Optical Products Division, Warren, New Jersey, is reported to have delivered sufficient energy per pulse at a repetition rate of 20 pulses per second to be suitable as an agile laser weapon. Over this wavelength range together with its doubled counterpart in the visible and near ultraviolet, protection against this laser will be difficult.

Another interesting LEL concept is the tunable free-electron laser (FEL). It may be assumed that low-energy tunable FELs are under development for electronic warfare purposes as short-distance antisensor weapons. It is possible to use more than one laser in an LEL anti-eye weapon system to make it more effective. This will certainly be costly, and careful study will be needed to determine if such a solution can be cost-effective, or if it is better to use one laser in the visible part of the spectrum and mass-produce it as a weapon. Such a less complicated anti-eye weapon will still be very agile if several laser wavelengths are used; in practice, it would be impossible to block the laser beam during combat due to the low transmission of any protective filter, which would prevent the soldier from seeing through the filter clearly and easily in a close combat situation. Also, not only must the filter be effective without blocking useful vision, but it should also be one that can be mass-produced at low cost.

The choice of a laser for an antisensor weapon is almost entirely dependent on the properties of the specific sensor chosen as the intended target. The most favorable situation is when the wavelength of the laser is within the operating wavelength range of the sensor, making it possible to attack and destroy the detector, which is the most sensitive part within the sensor system. If the sensor, like the eye, operates within the visible and near-infrared part of the spectrum and the wavelength chosen for the laser weapon is in the same region, it is possible to combine both antisensor and anti-eye effects in the same weapon.

The possible interaction between different lasers and different vulnerable military sensors is shown in Table 6.2. The examples of sensors given in the table are ones that are very sensitive to the lasers with which they are matched. Not all of the lasers used as examples are the best choices for an LEL weapon. They are used only to given an indication that possible sensors and lasers do exist in the visible and near-infrared portion of the spectrum.

An antisensor weapon may be designed to destroy or knock out enemy sensors for a certain period of time, thus preventing them from performing their intended tasks. The basic military requirement will certainly be to destroy, but that may be impossible in some situations and at times simply unnecessary. If the aim is only to disable the sensor for some time, a continuous-wave

TABLE 6.2.   Possible Antisensor Lasers

Spectral range	Lasers	Wavelength (nanometers)	Vulnerable sensors
Visible	Argon	514.5	Low-light TV
	Frequency doubled Nd:YAG	532	Image intensifiers
	Ruby	693.4	Range finder detectors
	Titanium–sapphire	660–1,160	Human eyes
	Alexandrite	700–815	
Near infrared	Gallium–arsenide	904	
	Nd:YAG	1,064	
	FEL	1,000–10,000	Thermal and IR missiles
Mid infrared	Deuterium–fluoride	3,000–3,800–5,000	
	Carbon monoxide	6,000	
	Carbon dioxide	10,600	Thermal detectors 8,000–12,000

(CW) laser or a laser with a high pulse repetition frequency has to be used at the proper blocking wavelength. In most cases, it requires less energy to disable temporarily than to destroy. An alternative use for a long-range HEL weapon may be to block the enemy sensors at longer ranges and then to destroy them if they manage to get closer.

LEL weapons may be of different sizes depending on the laser required. A very simple anti-eye laser operating in the visible spectrum at one or more wavelengths could easily be portable and also comparatively cheap. Such a small laser could be placed on a rifle and aligned with the normal sights to be used against enemy eyes in close combat. If the purpose is to counter or destroy electronic sensors at long ranges, then the laser with all its accessories will be a bulky and heavy system which must be carried on a combat vehicle; alternatively, a special close-combat laser vehicle, with the LEL as its main weapon, could be designed. It is possible to get an idea of how LEL weapons may look on the battlefield of the future by studying the weapons presently in the research and development phases.

SOME PRESENT LEL WEAPON SYSTEMS

Some examples of the research and development of LEL weapons have been published in the open literature, but, even with the benefit of this information, it is still difficult to give an altogether true and complete description of the present state of the art for these weapons. For obvious reasons, this is an area which is shrouded in very tight security. The following short summary of the information available is an attempt to put together the pieces of a rather fragmented picture of some of the ongoing research and development of LEL weapon applications. As mentioned before, the information is almost entirely from the Western world. Some ideas about the LEL weapon situation in the former Soviet Union are also given, but it must be pointed out that most of these conclusions are speculations based on second- or third-hand guesses and estimations.

Reliable proof currently exists concerning the fielding of a specific low-energy laser weapon with an anti-eye capability. At the very beginning of 1990, the British press revealed that a laser weapon called the Laser Dazzle Sight (LDS) had been in use on British ships for a number of years. The disclosure caused a public debate, raised many questions, and was discussed within the British Parliament. Subsequently, a lot of articles were published, and from them we can piece together some information about the properties of LDS, although very tight security regarding this weapon is still maintained by officials of the British Ministry of Defense. The purpose of the LDS is to blind pilots of attacking aircraft for such a long time that they are forced to abandon their attack before delivering their munitions. The requirement may even be to blind them for such an extended period of time that they must crash the aircraft or abandon it completely. The LDS could also be used against small boats attempting suicide attacks at larger ships during terrorist-type conflicts. The knowledge of the existence of such systems as the LDS might cause enemy pilots to fly higher, thus becoming better targets for air defense missiles and guns.

The LDS was developed jointly by the Ministry of Defense Royal Signals and Radar Establishment (RSRE) in Malvern and the Admiralty Research Establishment. Early trials were performed in 1981, and, according to some reports, several weapons were used during the 1982 Falklands conflict, resulting in the loss of three Argentinean aircraft. The LDS has since been used on British ships in the Persian Gulf and in other areas. The laser is contained in a rectangular section, barrel mounted on a tripod, and seems to be about five feet long. It may be manually aimed or assisted by a computer to lock on to its target. The LDS probably operates in the blue part of the spectrum and was originally developed from an industrial laser. It is obviously powered from the ship's electrical system, and the range should vary considerably with changing atmospheric and target properties. According to various reports, the range could be from 1 mile up to 3 miles. Even if the LDS system is primarily designed to make a pilot abandon his attack or miss a target, it can easily inflict serious eye damage and even blindness. It is not possible to only flash blind a person with a laser for a sufficient time in broad daylight without simultaneously causing permanent changes to his eyes. Temporary flash blinding by a laser is only possible when the eyes are more or less adapted to darkness.

The LDS should be seen as a complementary weapon to air defense missiles and antiaircraft guns. It will be useful at comparatively close ranges but will have no effect on aircraft delivering long-range standoff weapons. However, the importance of the LDS should not be underestimated. It has been fielded for a long time and has been thoroughly tested and evaluated. It will serve as a basis for the development of new LEL weapons in the United Kingdom by the RSRE for the army and air force.

Early in the 1980s, the U.S. Army initiated a development program for a battlefield antisensor close-combat laser assault weapon known as C-CLAW. The initial prototype was dubbed ROADRUNNER, and it was based on the combination of two lasers which could operate together at three different wavelengths. One of the lasers was a 1-kilowatt pulsed CO_2 device using a high pulse repetition frequency and a wavelength of 10,600 nanometers. The other was a Nd:YAG laser that could be used either at 1,060 nanometers or frequency doubled at 530 nanometers. Thus, ROADRUNNER could operate with one laser in the far infrared and the other laser either in the near infrared or the green part of the visible spectrum. The system was planned to be powerful enough to craze or fracture optical lenses, aircraft windshields, and other items made of glass or polycarbonate laminates.

The C-CLAW program had as its ultimate objective the development of an assault laser weapon to assist in offsetting the numerical superiority of the Warsaw Pact Forces in tanks and attack helicopters by neutralizing their sensor-dependent fire control systems. It was to be mounted on infantry armored vehicles and armed helicopters. The program was canceled in 1983 due to the cost and weight of the laser weapon. The original military requirement was for a weight of about 900 pounds, but, when the project was terminated, the laser weapon had already increased to 3,000 pounds.

When the ROADRUNNER project was made public, it was mentioned in press accounts that this new weapon could also attack the eyesight of enemy troops. This caused some debate in the media which reflected public concerns about the humanitarian issues involved, but, according to the Chief of Staff of the U.S. Army, the cancellation of the $212 million ROADRUNNER project at a later date had nothing to do with the adverse publicity. "Obviously, war is lethal," he said. "Even though C-CLAW can blind soldiers, it was not being developed for that purpose." However, in another press report, it was stated that among the reasons leading to the cancellation of the project was the extensive media coverage of the possibility that this system could also cause eye damage. It is difficult to judge what effect the media reaction ultimately had on the final decision, but it is significant to recognize that the idea of harming human eyes was not very well received by the public. Whatever the real reasons for cancellation were, the project had overrun both the cost and weight requirements, and the U.S. Army decided to turn to "a more promising technology" in the form of a new concept called STINGRAY.

STINGRAY is an optical and electro-optical countermeasure LEL weapon designed to detect and blind or destroy the enemy sensors that are essential for surveillance and for guidance of modern weapons on the battlefield. It is designed to be mounted on a Bradley M2 Infantry Fighting Vehicle, but it can also be used on conventional tanks and attack helicopters. According to another report, STINGRAY could be housed in long rectangular boxes mounted on the side of the Abrams M1 tank turret. STINGRAY can be operated in two modes. First, the laser may be used to detect enemy optics and electro-optics by tracing the small amounts of laser light that are reflected from the sensors when they are directed against the enemy positions. In this mode, the beam is used to scan the terrain. The operator sets the output beam on a wide angle to encompass the broader view and operates the laser at a minimum power level. In the other mode, the laser beam is highly concentrated to blind or destroy enemy periscopes, night vision equipment, and gun sights.

STINGRAY will be used as an add-on weapon on the Bradley Fighting Vehicle in conjunction with the ordinary 25-mm gun and TOW antitank missiles. The weapon is installed in a box on the opposite side of the turret from the TOW antitank missiles. The laser may be initially used as a laser radar, and, when the targets have been detected and identified, the Bradley crew can use the gun or the missiles against one target, while the STINGRAY laser blinds or destroys the sensors on a second target. The military requirement may even include the possibility of using the laser from a Bradley which is on the move against a target that also is moving. Not only would this allow the possibility of hitting a specific moving target, but it would ensure hitting a certain spot on that target, namely, the optic and electro-optic equipment. This can be done by directing the automatic flow of information from the radar mode to the beam or the aimer once the laser beam is narrowed and changed to the concentrated weapon mode.

As with all laser-based weapons, one of the main difficulties of STINGRAY is coping with the atmospheric conditions and the battlefield pollution. STINGRAY was designed with sufficient energy to hinder an enemy's target acquisition process if it is dependent on optical and electro-optical means. It is difficult to find a first-hand opinion on the probability of solving this problem, as the STINGRAY project is highly secret, and very little data have been made public. However, the project manager has answered a question concerning the possible successful resolution of the project's problems by retorting that "we wouldn't have this level of interest and support if we hadn't expected to be successful."

The STINGRAY program is sponsored by the U.S. Army Communications Electronics Command (CECOM), and the prime contractor, Martin Marietta Electronics Systems in Orlando, Florida, was awarded an initial contract in 1982. The General Electric Corporation is responsible for the development of the laser, which benefits from an innovative use of slab-shaped crystals, probably of neodymium:YAG, rather than the usual cylindrical rod. This configuration yields a tenfold improvement in beam quality by reducing heat-induced distortion of the lasing element. This solid-state slab laser will probably produce short pulses with high peaks of power in the near-infrared part of the spectrum. The U.S. Army considers STINGRAY a low-energy laser weapon which, according to the project manager, has a maximum power sufficient to temporarily upset or even wreck the enemy's optical devices.

The STINGRAY laser is installed in a stabilized mount in the turret of the Bradley tank. Its operation is assisted by a computer, and the laser gunner directs and fires the weapon from a visual display. The power is provided by a powerful alternator driven by the tank's main propulsion motor, which is a standard Cummins VTA-903T diesel developing 500 horsepower. This means that the ammunition available for the STINGRAY LEL weapon is almost unlimited as long as the fuel tanks are filled with diesel oil. The STINGRAY weapon itself is reported to be rather small in size and weighs only about 350 pounds.

STINGRAY has already reached a comparatively advanced stage in the research and development phase. A Bradley demonstrator that was equipped with a STINGRAY was field tested in 1986 at the White Sands Missile Range in New Mexico. According to an official statement from the Pentagon by Thomas P. Quinn, Principal Deputy Assistant Defense Secretary, "this system is now being considered for full-scale development." STINGRAY has been tested in a variety of atmospheric environments—wet, dry, dusty, and snowy. One of the major concerns is the possible interference of the operation of STINGRAY with the guidance of the TOW missiles when some of the laser energy is reflected from snowy surfaces.

The estimated cost for development of the STINGRAY is approximately $250 million, and it would take four to five years before mass production could start. The price is said to be between $500,000 and $1,000,000 per laser. The continuation of the project has been in doubt due to economic considerations,and in late 1988, when the Army showed its fiscal 1990 budget, STINGRAY had been canceled. However, funding was restored, and the Senate Armed Services Committee stated that development of the STINGRAY would enormously increase the combat capability of the Bradley Fighting Vehicle. The Committee even added sufficient money to the requested budget to begin full-scale development in the event that further tests confirmed the STINGRAY's value. Reports of recent field tests at Fort Bliss, Texas, have substantiated the effectiveness of the STINGRAY, and in October 1990 the U.S. Congressional Defense Authorization Conference approved its full-scale development. Six of the STINGRAY systems are scheduled to be delivered to the Army in 1994. The production of 48 STINGRAY laser weapons to be fitted to reconnaissance Bradleys in two armored cavalry regiments is currently up for review by the Defense Acquisition Board chaired by the Secretary for Defense Acquisition. Whether this means that all financial problems have been solved and that the project has a definite go-ahead all the way to field deployment still remains to be seen.

Though not designed specifically to damage the eyes of any enemy soldier exposed to the beam, STINGRAY would be powerful enough in its weapon mode for this task. However, the laser is reported to be too weak to injure the eyes of infantrymen when it is set to the laser radar mode searching for targets. According to a U.S. Department of Defense official, when the power level is increased and the beam is narrowed to concentrate its energy on an enemy sensor or soldier, it can "do tremendous damage." STINGRAY certainly has the potential to damage the eyes of enemy soldiers, both temporarily and permanently. Indeed, quite a few of the victims could be injured in a way that will render them legally blind for the rest of their lives. This may happen more or less by accident when the STINGRAY is used against sensors, but the possibility also exists that commanders might order STINGRAY type LEL weapons to be used systematically to damage the eyes of the enemy on the battlefield. The special advisor for armaments in the Pentagon, Dennis E. Kloske, during his congressional testimony in 1988 said that "the issue of blinding on a tactical basis is something everybody is going to have to deal with in the 1990s." A STINGRAY program critic in the Defense Department commented that the deliberate use of laser weapons to blind people is unethical and distasteful.

The U.S. Air Force is working on a LEL weapon based on the STINGRAY laser called the CORONET PRINCE. It is designed to protect aircraft against enemy air defense weapons which are dependent on optics and electro-optics for their effect. The crew will be alerted through the detection system in CORONET PRINCE that the aircraft is under attack or that an attack may be imminent. The crew can then identify the threat and counterattack using pulsed laser beams which will incapacitate the eyesight of air defense operators and disable the enemy's electro-optical sensors directing the weapons against the aircraft. CORONET PRINCE is intended to be pod-mounted, and its pointing, tracking, and environmental requirements differ significantly from those of the ground-based and vehicle-mounted STINGRAY. CORONET PRINCE will have to engage targets at longer distances and will, thus, require far more laser power than the STINGRAY. The current development of CORONET PRINCE is said to be somewhat behind that of the STINGRAY, although, according to an Air Force spokesman in 1988, "It is on track and on schedule, and we are looking at a full-scale development decision in 1989." The test model was nearing completion at Westinghouse's Baltimore facility and scheduled to undergo flight testing starting in early 1989, although details of the tests are not yet available. Future plans include an integrated electro-optic/radio frequency countermeasures program as well as integration of alternate transmitter technologies. The CORONET PRINCE as an advanced optical countermeasures pod (AOCM) will certainly increase the survivability of combat aircraft while penetrating enemy areas that are heavily defended by antiaircraft guns and missiles. Enemy systems which are dependent on their electro-optical sensors, beam riding technology, or visual tracking through optics will be more easily discovered and neutralized.

The U.S. Army has launched another STINGRAY-related LEL weapon for the protection of helicopters called the CAMEO BLUEJAY. It seems to be a 75-pound version of STINGRAY (300 pounds) designed to be mounted on an Apache attack helicopter. However, it is not expected to go into advanced development before the very late 1990s. The U.S. Air Force has yet another electronic warfare program called HAVE GLANCE, in which a pod-mounted, low-energy laser would be employed to confuse the heat-seeking function of infrared missiles.

The first LEL weapon projects that have any resemblance to the Flash Gordon idea of a hand-held ray gun are in the research and development phase in the United States. One project, the DAZER, is funded by the government and another, the COBRA, by industry. The DAZER is the first portable anti-eye laser weapon made public besides those antisensor systems which could also attack the eyes of the enemy. The DAZER uses an alexandrite laser and has been described by the Pentagon, in Senate Armed Services Committee testimony, as "a man-portable laser device for use by infantry to provide a soft kill against armored targets by attacking sensors, including television, night vision devices, and personnel." DAZER was initiated by the U.S. Army Infantry Center at Fort Benning. So far, the U.S. Army's Missile Command (MICOM) has directed the DAZER program, and the Allied Corporation's Military Laser Products Division, Westlake, California, has been given a contract to develop an LEL weapon based on their own commercial alexandrite laser. In 1981, the DAZER was put through a more or less successful, extensive series of tests at Fort Bliss, Texas, to measure its effectiveness against various targets.

The Allied Corporation produces and sells an alexandrite laser that is tunable from 700 nanometers to 815 nanometers with its maximum energy at 755 nanometers. One of the available rods has delivered 3.5 joules per pulse at a 20-hertz repetition rate, a total of 70 watts. Another rod, when Q-switched, delivered 600 microjoules per pulse in 33 nanoseconds and a peak power over 18 megawatts. Alexandrite is based on chromium atoms and is the synthetic analog of a mineral which was named after Czar Alexander I of Russia. The Allied Corporation manufactures the synthetic alexandrite under the trade name of Alexite.

It is not known for certain if the DAZER is based on the commercially available Alexite rods from Allied, but it may be presumed that a similar technology with a comparable energy capability is used. There are, at least, two important reasons for the choice of an alexandrite laser for an LEL weapon. The crystal itself tolerates higher operating temperatures than most other materials used in solid-state lasers, thereby reducing demand on the cooling system. Another important advantage of this type of laser is its tunability, which could make it very difficult for the enemy to protect sensors and eyes against an LEL weapon like the DAZER. The enemy can filter out many wavelengths easily, but any protective filter that covers the whole spectrum at once would have a low overall transmission and would not be useful in combat to protect eyes, as it would also block vision. This tuning capability will make the DAZER a formidable threat to badly protected or unprotected infantry units if it is systematically used against eyes instead of sensors.

The U.S. Army's MICOM has ordered two prototypes for advanced development under the terms of a $411,000 contract. The projected cost for one field unit is set at $50,000. DAZER is powered by a nickel–cadmium battery backpack, but a larger external power source may be used instead. The DAZER is submachine gun sized, weighs about 20 pounds, and is fired from the shoulder with a collapsing shoulder stock and a telescopic sight. The laser is presumably excited by a flash lamp. The device includes a cooling chamber and a pump to regulate its temperature, which is necessary even though the lasing medium can tolerate high temperatures. The basic beam is invisible, making it difficult for the operator to judge if the target was hit or missed. However, either frequency doubling or shifting could give a visible beam. Also a reference laser emitting in the visible part of the spectrum, such as a low-power HeNe or red diode laser, could be added.

The DAZER is a frequency-agile LEL weapon that is designed to attack very sensitive sensors mainly in the visible and near infrared part of the spectrum, including television, night vision devices, and the human eye. As all relevant data are, of course, secret, it is not possible to give any exact information on the weapon's capabilities. Supposedly, the DAZER will only be able to flash blind enemy soldiers temporarily, but the same press report also quotes a Pentagon R&D chief stating ambiguously that "it is possible to create a laser beam of sufficient intensity to permanently destroy the eye, and it is possible to create significant damage without destroying the eye permanently." As has been pointed out in previous chapters, it is not possible to flash blind a person in daylight with a laser without damaging the eye. If flash blindness can be made to last from 10 to 100 seconds, depending on the brightness of the flash and the illumination level, it is likely that the weapon will destroy eyes at close ranges and severely damage them at longer ranges. This will certainly be the case if the enemy is using magnifying optics.

The COBRA is being built by McDonnell Douglas Company

Electronic Systems Div., McLean, Virginia, and is in many respects similar to the DAZER. It is of the same size and shape and is driven by a battery pack. However, it does not use the same type of laser. The president of the company declined in an interview to discuss the wavelength, but he said that if the Army were to have a laser that could fire any of three different frequencies "you get to the point where the defense has to counter all three . . . then you can't look through optics, because the optics are black." This means that the COBRA is based on a single powerful solid-state laser operating at the same time at three different wavelengths, all in the visible part of the spectrum. The COBRA could be as devastating to human eyes as the DAZER. Although it is possible to some extent to have, at least, minimal protection against the DAZER by using a filter in the near infrared covering 700 to 815 nanometers, it presently seems impossible to have adequate protection against the COBRA.

Without a doubt, someday the DAZER and the COBRA will be very important weapons, as they are the first LEL weapons designed for use by the infantry against the sensors and eyes of enemy infantry. Furthermore, they have projected prices that will make it possible to produce and field them in large numbers. This development will set the standard for many other LEL weapons to come, and several other companies in the United States are reported to be working on similar weapon concepts. By the end of this century, weapons like the DAZER and COBRA will find their way into armies all over the world. The technology is not too sophisticated, and the laser industries that are already established in most countries will have the capability of developing similar LEL weapons. This is certainly the case within the former Soviet Union and countries that had signed the Warsaw Pact.

The former very tight security in the Soviet Union and the present unsettled conditions in the separated states make it impossible to give any thorough and completely true picture of what is going on in these countries as regards LEL weapons. Since 1981, the official U.S. picture of military developments in the Soviet Union has been reported annually in a document entitled *Soviet*

Military Power. The 1987 edition stated that "the Soviets have the technological capability to deploy low-power laser weapons, at least, for anti-personnel use and against soft targets such as sensors, canopies, and light material." Furthermore, it claimed that "recent Soviet irradiation of Free World manned surveillance of aircraft and ships could have caused serious eye damage to their personnel." These statements were followed in the next edition a year later by the following: "The Soviets are using their technological capability to move toward rapid deployment of low-power laser weapons with their military forces. Their tactical laser program has progressed to where battlefield laser weapons could soon be deployed." It is rather odd, against this background, that the reader of the 1989 edition will not find even one word about low-power laser weapons. This may mean that the authors did not find the LEL weapon important, or that they considered the former statements somewhat exaggerated, or that the whole subject had come under tighter security in the Soviet Union as well as in the United States.

The U.S. Army has published a new edition of a standard field manual covering the Soviet Army, FM 100-2-1. It is a detailed and unclassified review which contains a chapter on Soviet tactical directed energy weapons and warfare. The manual states that low-energy lasers are most likely to be used for blinding personnel and electro-optical devices, but it does not describe how a Soviet battlefield laser weapon would be configured. An article based on this new manual suggested that one Soviet LEL system is similar to the U.S. Army's own STINGRAY. The U.S. Army seems to have concluded, based on the extensive Soviet literature on the biological effects of laser radiation, that the Soviets are examining and working on the development of anti-eye laser weapons.

There have been additional reports in the open literature on Soviet development of LEL weapons. In connection with the previously discussed reports on the peacetime laser incident between Soviet ships and U.S. aircraft, it was said that the existence of Soviet laser weapons is generally known and that such weapons have been demonstrated and detected in various areas of the world. The Soviets have directed their research toward the practical side and have now reduced the size of their systems to a point where ground forces can be equipped with lasers for both offense and defense. Even if some experts such as David Isby seem very certain that the Soviets are more advanced in many laser weapon applications than the United States, that cannot be convincingly proved. The only definite conclusion that may be drawn is that there are no big differences between the countries in the former Soviet Union and the rest of the world when it comes to development of LEL weapon applications.

FUTURE LEL WEAPONS

The current research and development of LEL weapon projects indicates a very strong interest in this kind of weapon. Work is obviously proceeding in a number of countries, and we may be entering a new era of military weapon technology. LEL weapons have come to stay and will probably reach combat units in some countries during the mid-1990s. During the late 1990s and into the beginning of the next century, LEL weapons will be fielded in more countries and in increasing numbers. The two main areas of use on the future battlefield will be as antisensor weapons and anti-eye weapons. The design of dual-purpose LEL weapons combining these two functions will be possible, mainly with lasers emitting in the visible and near-infrared part of the spectrum. For obvious reasons, it is not possible to predict exactly which LEL weapons will be available on the battlefield within the next 10 to 20 years, but it is possible to describe several likely applications.

Airborne Weapons

Airborne LEL weapons will predominantly be designed for and used in the air-to-ground role. One of the main aims will be to detect, blind, and destroy enemy air defense installations. Other applications such as air-to-air combat between aircraft are possible but will probably not be as common as air-to-ground ones. It must be kept in mind that space and weight are always at a premium on modern military aircraft, and this severely limits the possibilities for internal laser weapons. This problem can be solved to some extent by a external laser pod; however, this slows the aircraft down and shortens its range, not always within acceptable limits for attack.

Four different possibilities exist for fielding airborne weapons: countermeasures against enemy air defense systems; as a part of anti-beam rider missile systems; directly against enemy homing missiles; and as an air-to-air assault weapon.

The first airborne LEL weapon concept suggested was a countermeasure system designed to protect the aircraft from many of the enemy battlefield air defense weapons. The main targets are sensors and sights used for fire control on antiaircraft gun and missile systems which operate in the visible or near-infrared part of the spectrum. A secondary target could be the personnel operating the sights and other electro-optical devices fitted to the different hostile air defense systems. The military requires a LEL weapon configuration that can detect and identify key sensors on defending antiaircraft weapons followed by automatic blinding or destruction. The LEL weapon has to be effective when the aircraft is on its final run toward the target, which means that the firing range has to be at least several kilometers. The hit probability has to be very high for a target on the ground or a ship. During the critical phase of the attack, it should be possible to blind simultaneously all sensors and sights that are sensitive to a laser beam. This is necessary if the aircraft is to press home an attack even in a heavily defended area. In most cases, this type of LEL countermeasure weapon will be pod-mounted and may be supported by laser radar for detection and identification of the target.

The second concept is an airborne anti-beam rider LEL weapon designed mainly to counter the increasing threat from antiaircraft missiles using laser beams to "ride" to the target.

Other air defense weapon systems that are dependent on using lasers in the fire control system may also be targets for this special type of laser weapon. It is of utmost importance that the anti-beam rider LEL weapon system be able to detect and immediately respond to incoming beam riding missiles and engage in successful air-to-ground combat. The aircraft has to be equipped with a high-technology laser detector which can sense the hostile laser radiation, calculate the position of the source, identify the threat, and automatically fire its own laser to blind or destroy the fire control system of the beam rider. If the missile is already on its way, there is no time for a manual response. Even if the laser beam is identified as that from a laser designator or range finder, a very fast reaction will be necessary. Existing beam riders, target designators, and range finders operate within the visible and throughout the near- and far-infrared regions. For example, the U.S. antiaircraft beam riding missile system ADATS uses a CO_2 laser operating in the far infrared, while the Swedish Rb 70 operates in the near infrared. An LEL weapon could detect, blind, or destroy targets operating at widely separated wavelengths, or could be tailor-made to counter one specific type of beam rider. It may be possible to combine the two concepts: countermeasures against both fire control sensors and those on beam riders. However, if the military requirements are set too high, costs will rise, and the system will become too complicated to operate in the field even with the support of high-technology threat analysis systems and powerful computers.

An airborne LEL weapon is a countermeasure system designed to cope with the threat from air-to-air or ground-to-air infrared homing missiles. Most fighters and other combat aircraft are armed with thermally guided infrared missiles to shoot down enemy aircraft. This is, of course, a major threat to any aircraft that meets the enemy in the air. Antiaircraft infrared guided missiles are used to protect combat forces and important installations on the ground or to protect combat ships at sea. An anti-infrared missile LEL weapon should blind, trick, or destroy the infrared target seeker before the missile closes in on the target.

The military requirements for such an LEL weapon system are as simple as they are demanding. It must detect and automatically blind the infrared missiles to force them to divert and miss the target. This has to be achieved even when the available time for reaction is extremely short, and the hostile missile may close in on the aircraft from any direction (in this case, a pod-mounted LEL weapon may be an appropriate solution).

The first three airborne LEL concepts mentioned above are mainly defensive ones. The fourth—an air-to-air assault laser weapon—is an offensive one. The targets are sensors and pilots in other aircraft engaged in air-to-air combat. This weapon system has to be powerful enough to engage the targets at rather long ranges under normal atmospheric conditions. The military requirements will be for a weapon that can detect and blind or destroy airborne electro-optical sensors used on enemy aircraft and on helicopters. However, the most important targets for this weapon may be the enemy pilots. It should be possible to blind a pilot, day or night, for the time necessary to force him to abandon the aircraft or crash. It should also be possible to operate this weapon independently or in conjunction with guns and missiles. The LEL weapon could be supported by conventional radar, a laser radar, or some other sensor system. It could be mounted internally or in a pod.

Shipborne Weapons

LEL weapons may also be designed to protect combat ships against threatening missiles and other "smart" munitions. As already mentioned in Chapter 5, concerning high-energy laser weapons, it is very easy to find a large enough space for energy sources aboard a ship. The main limitation for shipborne HEL weapons concerns the atmospheric problems that exist close to the water. One way to complement shipborne HEL weapons may be to add LEL weapons for use against targets with sensors highly sensitive to laser light. Three different concepts for shipborne LEL weapons are described below.

Sea-skimming missiles are probably the most dangerous threat to surface ships. Some of these missiles home in on the thermal signature of the ship using infrared sensors during the final phase of their attack. With this as a background, the first suggestion for a shipborne LEL weapon may be an anti-infrared missile laser. It is not only sea skimmers that use the infrared signature. Guided missiles and "smart" bombs may also be launched from aircraft and close in on the target for other directions than close to the surface. The target for this type of shipborne laser weapon will be the infrared sensors that guide many of the missiles that threaten ships. This countermeasure weapon should blind the infrared sensors for a sufficient time to force the hostile weapon to miss its target and crash harmlessly into the water. The LEL weapon system will have to detect, identify, and then automatically fire the laser against the target. This will certainly be a tough task, as the targets are both small and fast. The time factor is critical. Most sea skimmers do not open their infrared target seekers until they are comparatively close to the target. Thus, the usual military requirement for a quick reaction, high hit probability, and high kill probability will be difficult to fulfill. Also, it may be unnecessary to equip ships with LEL weapons against missiles at all if the development of HEL weapons succeeds. However, it will be possible to field LEL weapons for such use several years earlier and probably at a much lower overall cost.

Another shipborne LEL concept is a countermeasure to provide protection against low-light television sensors on missiles such as "smart" bombs. The shipborne laser weapon should be able to blind or destroy the low-light television camera. The military requirements for this weapon are very much the same as those for the LEL anti-infrared sensor laser described above. To use this weapon against low-light television cameras, it is also essential that the hostile target be detected and identified in time and that the LEL weapon automatically track and fire against the camera. The television camera has to be blinded as long as necessary to divert the missile from its intended target. Possibly, a second or two might be sufficient, but some missiles may require more time.

A third option is to use a shipborne air defense LEL weapon against attacking aircraft and helicopters in the same way as the British Laser Dazzle Sight (LDS). The targets would be sensors and pilots in enemy aircraft attacking the ship or otherwise passing close to it. It will be possible to engage targets only at distances of up to a mile or two. This means that aircraft launching long-range sea-skimming missiles or long-range standoff weapons will be out of reach for such laser weapons. The military requirement should be for an LEL weapon that could blind or destroy sensors and blind pilots at such far ranges that the enemy has to break off the attack before any weapons can even be used. Even in daytime, the effect of a laser beam on the pilot should be severe enough to force him to crash or leave the aircraft. He should be blinded for at least half a minute. Such a blinding weapon may not necessarily require complex and costly automatic operation. A high-energy laser beam will be more effective at longer ranges, but the disadvantages, cost, and complexity of such weapons have already been discussed at great length.

Ground-Based Weapons

Many more possible concepts for ground-based LEL weapons exist than for airborne or shipborne weapons. This trend is reflected in the large number of ongoing military projects for land warfare.

The first concept is a heavy combat assault weapon designed for the support and protection of combat units attacking or defending important targets. The weapon should be deployed at army brigade and division levels to attack or counter sights, optics, and electro-optic sensors fitted to enemy tanks, armored vehicles, antitank missile systems, equipment for forward observers, and helicopters. This laser weapon should disable all light-sensitive targets on the battlefield wherever they pose a threat to friendly forces. The military requirement should be for an armored combat

LEL weapon vehicle equipped with the LEL weapon as its primary weapon. It has to operate at long ranges independently of light conditions and should not be too restricted by the polluted atmosphere on the battlefield. The normal firing range should be no less than 1 to 2 miles, even during rather bad weather conditions. The LEL weapon system must be designed to detect, identify, blind, or destroy hostile sensors and sighting systems. It may be necessary to use more than one tunable laser to cover the parts of the visible and infrared spectral range of interest. The weapon should be powerful enough to craze glass at average battlefield combat ranges. This weapon certainly has to be a very frequency agile device. Such complex military requirements will also mean that this will be a very costly weapon system to develop, manufacture, and maintain within the armed forces. Also, if it is to be really cost-effective, large numbers must be fielded.

The second alternative is not as ambitious as the first one even though the two concepts have many similarities. In this scenario, an LEL weapon would be used as an add-on combat assault weapon to complement the conventional main weapons. The laser weapon should be used against the same types of targets as the main weapon. The basic idea is to field a simpler and, thus, cheaper LEL weapon family that could be deployed in greater numbers. Such a weapon could not cover a very broad range of wavelengths, and the design may have to be limited to one small part of the spectrum. The military requirement should be for a weapon that could be fitted to tanks, other armored vehicles, and weapon systems such as antitank missiles. It has to be able to detect and blind one secondary target, at least, and hopefully more, while the primary target is engaged by the main gun or missile. Selection of the wavelengths will be based on the actual sensors and sights used by the intended target. In most cases, it may be necessary to use a tunable laser or alternatively to use more than one laser in this LEL weapon application.

A third concept is an antilaser LEL weapon designed to be used against other laser-based devices such as range finders, target designators, and beam riding missile systems. Such a weapon may even be used against other laser weapons.

An important prerequisite for an antilaser weapon is a very efficient laser detector that immediately gives the correct position and identity of the enemy laser source. If the threat is identified as coming from a beam riding missile system or from a target designator, the time factor is critical, and it is necessary to fire the antilaser weapon automatically against the most threatening laser source. The designer has to decide which laser is the main threat and then adjust the range and wavelength accordingly. It is not possible to cover all the wavelengths of threatening lasers in a single weapon without making it complicated, bulky, and expensive. The military requirements are, to a great extent, the same as for the corresponding airborne LEL weapon.

An antiaircraft LEL weapon is a fourth concept. It should be used against sensors and crews on attacking aircraft and helicopters. Such a countermeasure weapon could be a very cost-effective supplement to conventional antiaircraft guns and missiles. The requirement in this case is for a weapon that can detect, blind, or destroy sensors and human eyes aboard the aircraft. This has to be done at such a long range that the attacking aircraft or helicopter is forced to break off the attack before its weapons are launched. The laser should be powerful enough to blind the pilot, day or night, for such a long time that he has to leave his aircraft or is forced to crash his helicopter. This is easier to do at long ranges at night when the eyes of the pilot are dark adapted and thus extremely sensitive to the bright laser light. The incoming targets could be detected by a conventional or laser type of radar. It should be possible to operate the laser in either an automatic or a manual mode. Even the two other ground-based laser weapons mentioned above could have an air defense role.

Finally, helicopters could be equipped with countermeasure LEL weapon systems against lasers, beam riders, and infrared missiles in much the same way as aircraft. Another possibility is to turn the helicopter into a combat assault weapon for support and protection of brigades and divisions. A helicopterborne weapon could use the same basic ideas as the first and second alternatives for the ground-based combat assault weapons mentioned above.

Anti-Eye Weapons

The LEL weapon concepts described so far have mainly been in the category of antimatériel weapons. In some cases, the military may require, as a secondary task, that these weapons be capable of blinding the eyes of the crew on a tank or in an aircraft. This is, of course, an antipersonnel use, but the main aim is still to destroy the aircraft by forcing it to crash or to destroy the blinded tank with a gun or a missile. In this case, the eyes are more or less secondary targets. Anti-eye LEL weapons are quite different. The sole target of these controversial weapons is the eyes of the enemy soldiers, either naked or behind magnifying optics. The ultimate aim is to injure or destroy the eyes sufficiently to stop the soldier from fighting that war or any other war. This means that the soldier has to be immediately blinded, and the injury must damage the eyes to such an extent that the soldier cannot recover and take part in the fighting again. The greater the number of ocular injuries inflicted on the soldiers and the more severe these injuries, the bigger will be the burden to the medical facilities and the society of the enemy. Another important effect caused by a great number of blinded or severely injured soldiers will be a psychological one: the will of the soldier to keep on fighting may be diminished by the existence of laser weapons on the battlefield.

The military requirement for an anti-eye laser weapon is that it be used systematically to flash blind, injure, and destroy the eyes of enemy soldiers in combat at ranges shorter than a mile. The weapon should be small, light, hand-held, battery-powered, and very frequency agile within the retinal hazard region. It has to be designed for mass production and should be as cheap as a machine gun or a rifle. It could be designed as an independent weapon or as a laser device to clip on ordinary rifles, machine guns, and antitank weapons. One of the important requirements is that the weapon be cheap so that it can be distributed to combat units down to the squad level or even to individual infantrymen. The wavelength may either be tunable or chosen in a part of the spectrum where, due to lack of overall filter transmission in the visible region of the spectrum, it is impossible to have any protection.

The role of the laser as an infantry weapon needs some clarification. The use of the term blindness to describe the effect of an anti-eye laser weapon is somewhat misleading. It is true that, in most cases, there will be visual loss sufficient to constitute legal blindness, but this is far from total blindness. In the overwhelming majority of cases, the affected individual will still be able to get around without any assistance, but, as has been indicated, precision tasks, such as driving a motor vehicle at any rate of speed, will be difficult or impossible. However, the degree of visual impairment or interference with mobility is the crucial point of the battlefield. It will be impossible for the soldier with serious eye injuries to continue fighting. The most important use for the laser by the infantry will be at longer ranges to cause casualties and affect morale, but, as an assault or antiassault weapon, the LEL laser will be essentially useless. That is, an antipersonnel laser weapon will probably be of no assistance during hand-to-hand combat and will, in fact, probably not even act as a significant deterrent to close combat. Although certain flash blinding laser systems may be helpful in this regard, they will probably be less effective than a large flashbulb. Such disorienting devices could not generally be used on the battlefield in the daytime but could be of some use at night; however, the fact that they would have similar visual effects on friendly personnel as on the target or enemy personnel would tend to minimize even such help.

In assessing the LEL laser as an infantry weapon, it must be remembered that such a weapon could only be used against targets very sensitive to laser light—in this case, the eyes or other laser-sensitive sensors. Furthermore, the LEL laser weapon is of no value for penetrating other parts of the human body, nor can it go through any protection such as body armor, vehicles, or military equipment. This means that the LEL weapon cannot replace the rifle or machine gun. It could be a complementary weapon clipped on to the rifle or used as a separate weapon while other infantrymen use conventional weapons.

There are many targets on the battlefield for such anti-eye weapons, including dismounted infantry acting in forward artillery controller teams, forward air controller teams, surveillance teams, commanders, and many other individual soldiers or teams. In fact, every soldier looking through a pair of binoculars, sights, or using his naked eyes to look in the direction of the enemy is at risk. This is even more so during night. The eyes are dark adapted and more sensitive, and it is easy to trick soldiers to look in a certain direction by showing a small light. What have been mentioned are only examples. There are many more situations in which a silent, more or less invisible anti-eye laser weapon will be of good use to the infantry soldier.

To summarize, the major use of lasers as antipersonnel weapons will be at ranges over 20 yards to cause casualties as well as destroy morale, as this is a deterrent for enemy movement or preparations for assault. It is unlikely to sufficiently disable an assault team to prevent it from accomplishing its goal within a 20-yard range.

CONCLUSIONS

A review of LEL weapons for use in the air, at sea, and on land has been presented, giving specific examples of what may be currently considered by a military staff and in military research and development. The list presented is certainly not complete, and there may be many more possibilities in addition to those which have been described. Many LEL weapons will reach combat units during the 1990s, and the number will certainly increase after the year 2000. Only three events can stop or even slow the process—a worldwide disarmament, a very successful technological development of effective and cheap protection, or an international ban on anti-eye laser weapons.

SEVEN

Protection and Countermeasures

It seems inevitable that the battlefield laser threat will markedly increase in the coming years. This will be as a result of not only the development and implementation of laser weapons but also the increasing number of other helpful laser-powered devices such as range finders and target designators. Therefore, it will be necessary for armies to protect their sensors and personnel by introducing passive as well as active countermeasures for laser technology. The primary laser threat will come from laser weapons, although conventional weapons guided to their targets by lasers will also constitute an indirect laser threat, as will be demonstrated later in this chapter.

Protection and countermeasures against laser weapons are difficult problems which so far have remained unsolved despite years of research. A simple and cheap eye protection against anti-eye laser weapons still does not exist; consequently, protection of personnel involves many complicated factors ranging from filters to defensive battlefield behavior. This chapter will mainly deal with what we can do to protect personnel, sensors, and combat units against the laser beams from low-energy laser (LEL) weapons. Protective measures required to counter high-energy laser (HEL) weapons will only be described briefly.

It is, of course, possible to stop any laser beam in its path before it reaches the sensor or the eye, but stopping the beam may mean that neither the sensor nor the soldier will be able to see and register crucial information from the environment, as stopping the laser beam more or less prevents viewing. As long as the sensor or the eye must receive some of the visible or invisible radiation from the electromagnetic spectrum, it will also be exposed to any incoming laser energy at exactly the same wavelengths. When military planners and research institutions investigate the possibilities of defense against the increasing battlefield laser threats, they are faced with three main possibilities. The first one is to block the laser beam before it reaches its intended target. The second way is to use tactical means and countermeasures, and the third is to change the visual behavior of soldiers and combat units.

BLOCKING THE BEAM

If it were possible to block the beam completely from reaching the sensitive parts of electronic sensors or the eyes of soldiers, this would obviously eliminate any hazard, but this alternative would only be useful if the sensors or eyes are not forced to abandon their assigned tasks at the same time as they are being protected. There are some widely different ways to accomplish this. The beam may be blocked by an attenuating filter placed in front of the sensor or eye. Such a filter has to be effective against the specific laser wavelengths that are being used against it, yet still allow sufficient amounts of the remaining wavelengths to be transmitted to the sensor or eye to be used for target detection or general surveillance. Another obvious way to block the beam from reaching the eyes is to use indirect viewing methods through an electro-optical system such as a thermal sight or a low-light television device. When the hostile laser beam hits the electro-optic system, the laser energy is prevented from reaching the eyes. The electro-optic system detectors may burn out, and the electronic eye will be blinded, but, with proper design, it will be possible to replace the destroyed parts and then use the sensor again. Other possible ways of blocking the beam that have to be investigated further are the use of veiling smoke, rapidly closing and opening shutters, and the rather odd and primitive black eye patch. The black eye patch stops the laser beam from reaching one of the eyes, which is thus protected but is at risk for later damage. All of these methods of blocking the beam will be described in more detail in the following sections.

Filters and Shutters

Lasers are extensively used for many different purposes outside of the military, in industry and medicine. Protection of personnel is a prerequisite to many uses of lasers in these fields, and there are special filters available as well as a whole array of other protective and control measures. These are fully described in the literature. However, only a few of these basic types of filters are rugged or versatile enough to be used in military operations. Available filters do not really have the necessary properties to solve the problems created by battlefield lasers and laser weapons, as they only give protection against a small number of fairly discrete laser wavelengths.

Anyone who wears sunglasses is using filter technology to protect his eyes against the light. For example, a pair of gray sunglasses blocks light of all colors almost equally well but it is not dark enough to give protection against laser light. If the aim of sunglasses is to block all light to a certain degree, the aim of a laser light filter is to allow as much light as possible to pass through except at the wavelengths that include the hostile laser light. The first laser protective eye wear was developed in 1962 by Dr. Harold Straub of the U.S. Army Harry Diamond Laboratory. He installed a blue-green glass filter plate (Schott BG-18) into a standard acetylene welding goggle frame.

When describing the level of protection obtained from a filter against a given laser wavelength, the term optical density is normally used. The optical density is a measure of the attenuation or weakening of the light beam afforded by a certain thickness of any transmitting medium and is logarithmic in nature, as described by the following mathematical expression:

$$OD = \log(I_0/I) \tag{7.1}$$

where OD is the optical density, I_0 is the power of the incident beam, and I is the power of the transmitted beam. This means that a filter with an optical density of 3.0 attenuates a beam passing through it by a factor of 1,000 or to 0.1% of its former intensity, and a filter with an OD of 6.0 attenuates by a factor of 1,000,000 times or to 0.0001%. When two highly absorbing filters are stacked, the optical density is essentially the sum of the two individual densities at any particular wavelength. The values for optical density corresponding to various percent transmission values are given in Table 7.1.

Sunglasses normally have an optical density of less than 1. For laser filters, the requirements are much higher, and researchers

TABLE 7.1.   Optical Density and Percent Transmission[a]

Percent transmission	Optical density	Percent transmission	Optical density
100	0.00	0.1	3.00
50	0.30	0.01	4.00
10	1.00	0.001	5.00
5.	1.30	0.0001	6.00
1.	2.00	0.00001	7.00
0.5	2.30	0.000001	8.00

[a]See Eq. (7.1) for the relationship between optical density and percent transmission and associated text for discussion of terminology.

look for optical densities as high as 18. As pointed out earlier, it is necessary that the laser protective filters transmit as much visible light as possible. A low transmittance of visible light means either that the soldier will not get crucial visual information upon which his life may depend or that he will experience severe eye fatigue. The problem is similar for sensors and even worse for aviators, who not only have to look for the enemy but also must see well enough to fly the aircraft.

Colored laser filters may make it impossible to see colored signals such as red warning lights, the illuminated red marks inside sights or traffic control signals. Certain color filters will affect daylight vision differently from night vision, since scotopic (night) and photopic (daylight) vision use different spectral regions. Thus, blue-green filters are better for night vision than red or orange filters and vice versa. The names of filters are sometimes confusing, but confusion can be minimized by remembering that the filter color is the spectral region that is less attenuated. A blue-green filter is transparent for blue and green but blocks out red and yellow.

It is difficult to determine exactly what the minimum acceptable transmission for filters used by combat units should be, since this value always depends upon the specific situation, the laser threat, and the assigned task for the unit involved. However, for most of the eye protection used on the battlefield, the overall transmission of visible light by the filter should not be less than 80%.

So far, the most widely used filter for military applications is the absorption filter. It may be of almost any color and is manufactured out of glass or plastic. It works simply by absorbing most of the light at one or more specific wavelengths. However, as these unwanted wavelengths are absorbed, these filters themselves can be damaged by powerful laser beams as a result of heating. A serious drawback of absorption type filters is the lack of an ability to shift from strong absorption at the laser wavelength to very weak absorption for nearby wavelengths. That is, they do not have a sharp transmission "cutoff" or "notch" near the laser wavelength in the visible.

Absorbing dyes can be used in an absorption filter. Such dyes may have a high optical density, often as much as 16 to 20, and can easily be combined with a plastic substrate. Although the dye absorption filters provide a simple, lightweight, and cheap solution, they are not problem-free. Their high optical density may be lost by aging, oxidation, or exposure to sunlight. This latter effect is called solarization, and, in one case, a decrease from a density of 4.0 down to 1.2 was found in just three weeks with daily exposure to sunlight. Absorbing plastic filter materials are easily scratched, the surface may be damaged by chemical solvents, and quality control during manufacture appears to be difficult.

An alternative to plastic absorption filters is colored glass, which has been used in laboratories for laser protection in the visible part of the spectrum for years. Inorganic colorants in glass are quite stable, and such filters are the simplest to produce. It is also relatively easy to make them resistant to mechanical wear and damage from very intense laser sources. However, in the near infrared, only a few different wavelength absorption curves are available.

A different type of filter is derived from the insulating or dielectric coatings developed to suppress reflections from lenses. Thus, antireflective coating principles have been used in the design of the modern dielectric interference filter, which selectively reflects different wavelengths and which depends on the effects caused by multilayer coatings. Such filters consist of many alternating layers of different dielectric materials. One of the great advantages of such filters is the possibility of a relatively sharp spectral "notch." This means that they can selectively reflect a given wavelength while transmitting to a high degree at other nearby wavelengths in the visible part of the spectrum. The designer can control transmission and reflection as a function of wavelength. As these filters reflect rather than absorb the light, they can withstand more laser power than absorption filters, but they have one great flaw. The color of the light reflected changes as the viewing angle changes. This means that the effectiveness of an interference filter for any particular laser depends on the angle of incidence of the laser beam. As the angle at which the laser beam strikes the surface of the filter changes away from the vertical, the wavelength blocked by the filter shifts. Some filters even become almost transparent at the design wavelength if the laser beam hitting the filter is only 20 degrees away from the perpendicular, which, of course, is unacceptable for outdoor applications. This effect can be demonstrated by looking at an interference coating with white light. As the filter is tilted, thus changing the viewing angle, the color seen will change. For this reason, interference filters are mostly found as coatings applied to absorption filters. The combination is primarily used in optical systems with a small field of view, where the range of angles at which light hits the filter is limited.

Instead of depositing the dielectric layers one by one on each transparent plate to make filters, a hologram can be made of one plate, and it will act on light just as the original multilayer dielectric filter did. The hologram can be reproduced mechanically by pressing plastics, just as a compact disc is made. The holographic filter has the same basic advantage as the dielectric coatings in interference filters, as it provides narrow-notch wavelength protection. However, the holographic filter also has the same disadvantage of dependence on the angle of incidence of the laser beam as the multilayer interference type.

In addition to wavelength control, holographic technology offers the filter designer a method of scattering or bending the laser light in a controllable way, thus changing the image to a different size or moving it to a new location in the optical system or eye. So far, holographic filters are expensive to produce, but potentially they can be made cheaply in large numbers using plastic molding technologies. Research is now in progress with the aim of making it possible to use this technology on a large-scale basis for the protection of soldiers in the field.

Filter Deployment in Active Service

There are several different ways to implement laser protection in combat units. The crews riding inside tanks or other armored vehicles can be protected through filters and coatings applied to sights and vision blocks. The chief objective is to stop the laser beam from penetrating into the vehicle or its sensor systems. If the protection is designed according to this principle, armored vehicle crews will not require individual eye protection as long as they are inside the vehicle.

Flight crews of airplanes and helicopters can be equipped with laser protective visors. However, the possibility of turning the windshield or the cockpit cover (the canopy) itself into a laser filter by using coatings or some other method does not seem realistic yet. The need for laser protection of flight crews is always in conflict with the requirements for unlimited visibility in day and night combat situations, as aviators can tolerate far less light loss than other soldiers. Some safety regulations even state that laser eye protectors are not normally recommended because of loss of peripheral vision due to blockage by the frame, reduced visual sensitivity, and degraded color contrast. To solve the problem, at least to some degree, some armed forces are looking into the development of wraparound laser protective visors which can be mounted on the helmet. The U.S. Army has reported that such panoramic visors will be used to protect against several possible enemy lasers and at the same time minimize the light loss necessary to provide this protection. To maximize the protection during both day and night, it could be necessary to have different visors for different light conditions. However, even if such visors are accepted and fielded, they will probably not give sufficient protection against any LEL weapons that can use varying wavelengths.

Protective eye wear for infantry soldiers has been supplied to combat units in some countries. Since 1988, the U.S. Army has furnished Ballistic and Laser Protective Spectacles (BLPS) to high-priority units in the Army and the Marine Corps. The original $3.9 million contract for 100,000 pairs was regarded as a "quick fix" solution until a follow-up development could offer increased acceptability and protection against additional threats. The BLPS are dye-filled polycarbonate plastic filters which will protect eyesight against the low-energy lasers most likely to be encountered on today's battlefield, specifically the two or three wavelengths used by common range finders and target designators based on Nd:YAG and ruby lasers. It has been reported that these polycarbonate spectacles incorporate dyes which remove the energy of only two wavelengths. These spectacles will not give any protection whatsoever against frequency-agile LEL weapons. They will only protect against a small fraction of the visible and near-infrared parts of the spectrum, and, since they are a passive countermeasure, they must be worn for long periods of time in case of a laser attack. One problem may be the reluctance of the soldier to accept any reduction in his combat effectiveness from reduced vision and the possibility that the spectacles or goggles may be damaged or lost. However, the mechanical protection against flying particles that the spectacles also offer is an additional and major advantage. The polycarbonate ballistic lens reduces the probability of eye injury from fragments by approximately 50%. It may be concluded that present technology is not sufficient to provide complete laser protection against LEL weapons to tank crews, flight crews, and infantry soldiers. It is even questionable if it is possible to have total protection against some of the powerful designators and range finders present on the battlefield today. Thus, it is unquestionably necessary to investigate the technological possibilities of developing new and much more efficient filter or shutter eye and sensor protection.

Future Developments in Protective Equipment

Military services in many countries are making vigorous efforts to find ways to protect soldiers and sensors against laser radiation. One future possibility is the use of nonlinear optical polymer materials which change their optical properties in the presence of intense lights or electric fields. They undergo a rapid molecular change of polarization and can become nearly opaque to laser radiation during the time they are hit with laser radiation and then return to the clear state when the laser pulse ends. Thus, these filters could block the laser pulse but would allow ordinary but less bright light through almost unchanged. It would probably be necessary to focus the incoming laser beam to initiate the necessary changes in the sensitive polymer material. This means that small focusing lenses would have to be incorporated into goggles and visors. Another way to block the damaging laser light may be to use filters that discriminate between coherent and ordinary light and let ordinary light pass for normal vision. These ideas have a high theoretical potential, but the problems involved are very complex. It will certainly take several additional years before any of these ideas can be assessed as realistic or not. Several other technologies that can be used to block laser beams are currently being investigated, such as optical switches and liquid crystals. Mechanical shutters that can be triggered to close the optical pathway before the light reaches the eye or sensor are not realistic possibilities since it does not seem possible to have a reaction fast enough to compete with the laser beam moving at the speed of light, and the first group of photons that reach the target before the shutter closes may be devastating enough to destroy the eye or sensor. Ultimately, problems with the mechanical approach seem impossible to solve.

One way to provide better protection for sensors is to laser harden them when they are designed. The detectors within the sensors may be designed to be easily replaceable, and unnecessary sensitivity to laser light might be avoided by a careful choice of materials. The main problem is that any filter protection can be defeated or bypassed if the enemy uses light of a different color or wavelength or simply uses a more powerful laser. Thus, the protection has to be independent of wavelength and power if the threat from LEL weapons is to be successfully averted. The possibility of developing and fielding laser eye protection that will give the infantry soldier a sufficient level of protection seems very small. Not only is it necessary to make such devices cheap enough to be bought in huge numbers, but they must also be so easy to wear that the soldier can have them on day and night, even in life-threatening situations preparing for close combat.

Indirect Viewing

Indirect viewing is an interesting possible way of protecting the eyes of crew members in high-value targets such as tanks, helicopters, and aircraft. If a television system, a thermal imager, or a light intensifier is used for observing the battlefield, tracking, and firing guns, the crew members do not have to view the battle area and the enemy directly with their own eyes. If a hostile LEL weapon is fired, only the light-sensitive parts within the electro-optical devices will be destroyed or blinded. However, the disadvantages are obvious. Such systems are very expensive and will only be worthwhile for the protection of high-value weapon systems. It is not possible to use them to protect infantrymen and other soldiers on the battlefield. Even for the protected few, there are drawbacks. It is difficult in a combat situation, which is highly stressful and often life-threatening, to be unable to observe the battlefield with the naked eye. If the enemy blinds the equipment used for indirect viewing, the soldier must either break off the action and take cover, go on fighting using his unprotected eyes, or try to repair the equipment during the stress of a hostile engagement.

Smoke

Smoke can be used as a countermeasure for two purposes. It makes target acquisition difficult for the enemy, and it can also diffuse and absorb the laser beam, making it less dangerous to sensors and soldiers. Smoke is frequently present on the battlefield from exploding munitions or burning vehicles, trees, and buildings. Its effect is mostly to limit the possibility of observing the targets and accurately directing weapons. However, some electro-optical sensors, such as thermal sights, can see through the smoke and haze and give the gunner a possibility of hitting the enemy without being seen himself.

It is necessary to use the right type of smoke if it is intended to give any protection against a laser threat. The effects of the smoke are dependent upon the size of the smoke particles and the wavelength of the laser. As the size of the particles approaches the wavelength of the laser, more of the energy will be scattered than attenuated. The scattering is very wavelength dependent and is most effective in the blue. The interaction between different types of smoke and the wavelengths used by anti-eye laser weapons is of great interest. The depth of smoke, in yards, required to reduce laser transmittance to 1% of the incoming energy is given in Table 7.2 for several types of smoke. It should be noted, however, that smoke protection even to the 1% level is comparable to a filter with an optical density of 2.0 in laser attenuation, which is not usually sufficient to be very useful on the battlefield. To be really effective against a laser, the smoke layer may have to be 100–1,000 times more dense.

If a combat unit desires to limit a laser threat with smoke, a large area of coverage is required in order to provide continuous protection as opposed to localized grenade-generated smoke. This can be achieved by smoke generators deployed within the forward combat units. When the smoke screen has been developed, it may cover the soldiers and protect them, at least to some extent. However, the use of smoke is certainly not without drawbacks. The smoke will interfere with the use of many types of electro-optics and other means for observing the battlefield. Overall, combat performance will be lowered due to loss of unaided vision, and control of the battle situation will be harder to achieve. Maintaining smoke for long periods of time over vast areas will require tremendous supplies and extensive logistics to feed the smoke generators. The use of smoke is very dependent on weather conditions such as wind and humidity. One of the main disadvantages of smoke is the time it takes before a smoke screen of the necessary density can be placed in the right position. It will be too late to use smoke for protection after the enemy has initiated the

TABLE 7.2.   Relative Transmittance and Optical Density for Various
Fog or Smoke Compounds in the Visible and Near Infrared[a]

Compound	Smoke depth[b] (yard)		Optical density	
	500 nm	1,060 nm	500 nm	1,060 nm
White phosphorus	15	23	0.9	0.7
Hexachloroethane	16	18	1	0.8
Fog oil	22	34	0.7	0.5
Titanium tetrachloride	10	23	1.1	0.7
Naphthalene	13	29	0.9	0.6
Anthracene	13	32	0.9	0.6
Sulfur trioxide	4	9	1.4	1.1
Silicon tetrachloride	9	17	1.1	0.8

[a]Transmission distances and optical density of the compounds are for standard methods of generation.
[b]Depth or thickness of smoke in yards required to reduce laser transmittance to about 1% of incoming energy is given at selected wavelengths in the visible and infrared. It should be noted that lasers using wavelengths outside the retinal hazard region require much more smoke to reduce the transmittance.

use of anti-eye laser weapons. Some laser weapons may even be powerful enough to penetrate through the smoke and cause eye damage anyway. Smoke is only one of several means to cope with an antipersonnel laser threat, and its use should be actively integrated into the laser protection protocol. In some situations, smoke can make a substantial contribution to the overall protection, but, in most tactical situations, its use will be of minor aid, and, even then, both the commanders and the soldiers must be aware of its limitations.

Black Patch

When the infantry and other combat soldiers are ordered to fight in an area where the laser threat has proven to be substantial, and if no other means of protection exist, a black patch over one of the eyes may be a solution. This idea, originally suggested by the military, of course seems odd, but such a measure would, at least, save one eye if the soldier is the victim of a hostile laser beam.

Obviously, this method has a lot of disadvantages. The soldiers will lose their depth perception to some extent and will not be able to use optics properly, since they are mostly designed for two eyes. The soldier's field of view will be significantly reduced, and it will be more difficult to perform normal tasks and move around quickly, especially at night. The psychological effects are unknown but undoubtedly could be substantial. The black patch will perhaps increase the soldier's fear of the laser threat instead of abating it. However, if the infantry has to fight, and the enemy is using a laser in the anti-eye mode, the black patch may very well be the one and only way to cope with a very difficult situation.

ANTI-LASER WEAPONS

The rapid development of laser weapons will certainly spur active countermeasures such as specific antilaser weapons. Today, many armed forces have antiradar missiles for use against active enemy radar stations. Such missiles are mostly airborne and have a considerable range. It is likely that missiles designed to home in on laser weapons, laser range finders, laser designators, and other laser devices will also be developed and fielded during the latter part of the 1990s. In order to be able to counter a laser that only emits pulses for a very short period of time, the missile has to "remember" the position of the hostile laser and home in on that specific position. Antilaser missiles could potentially make it unsafe to keep laser beams on for any significant duration. Such an antilaser missile will obviously be expensive and should, therefore, be used against the most threatening laser sources on the battlefield. This may mean that enemy laser weapons and enemy lasers that are directing conventional munitions should be given priority. Development of antilaser weapons would certainly also lead to the development of a variety of countermeasures, including laser decoys to trick the antilaser missiles and divert them in the wrong direction.

As a counter weapon against laser activity, the laser has many advantages over the missile. The most important is the speed with which an antilaser laser weapon can react to a laser threat. As antilaser weapons, lasers will probably be more frequent and will perform more of the required tasks on the future battlefield than missiles. However, both ways of coping with the laser threat will still be used. Since a missile can carry more energy and destroy more of the target than the enemy laser source or enemy sensor, it will be a necessary complement to the antilaser weapon.

When a hostile laser beam hits a target, retroreflection could be sued to send the laser beam back where it came from. Corner cube reflectors familiar to many from traffic control signs and roadway markers have the property of sending any reflection back in the exact direction from which it came. Corner cubes are used today on targets in laser simulation systems to reflect the very weak beams used in these systems so as to indicate target acquisition. For obvious reasons, these reflected beams have to be harmless. If the powerful beams from an LEL weapon or from range finders and designators were reflected back to the laser source, the laser device could be destroyed, and the personnel operating the laser would be injured. This suggestion has some drawbacks, but it would be possible to give protection against the specific wavelength the weapon emits. Also, a retroreflector will certainly give away the position of friendly units to enemy laser radar. However, a cautious use of retroreflectors on some dummy targets or at other positions in the area where the enemy is aiming their lasers may place some limitations on the laser threat as a whole.

LASER DETECTORS AND ALARMS

Most of the methods of dealing with a laser threat involve some active countermeasures, evasive maneuvers, or direct engagement. Generally, these methods will be used by combat units in various combinations. However, all of these alternatives presume that, in most situations, the hostile laser has been detected, identified, and exactly localized within the extremely short span of time available. Crews in fighting vehicles, airplanes, and helicopters, infantry soldiers, and others must be instantaneously warned of specific laser threats. Thus, it has become necessary to develop and field small, rugged, lightweight, and inexpensive laser warning systems capable of detecting and determining the direction of incident laser radiation with a comparatively high degree of accuracy.

Work on laser detectors has been going on for some time in many countries. A first generation of laser detectors has already been fielded. One example of such a detector was presented by Yugoslavia in 1988. It is a device that can detect laser radiation at wavelengths between 660 nanometers and 1,100 nanometers within a horizontal angle of view or sector of 360 degrees, but within only a rather narrow vertical sector. The direction of the hostile laser is identified as being in one of 24 sectors, each with a width of 15 degrees, and the warning will be a loud noise or other acoustic alarm signal. However, this particular detector does not provide very specific information on the direction of the hostile laser. A German company, MBB, has worked since 1983 on a second-generation laser detector, called the Common Optoelectronic Laser Detection System (COLDS), which is sensitive to all lasers in the visible and near-infrared range. The COLDS looks very much like an insect's multifaceted eye, with about 200 lenses, each of which is joined to a detector by an optical fiber. Such a system can give a very precise direction to the origin of the laser that is detected. The British company Ferranti sells a laser warner for aircraft, helicopters, and armored fighting vehicles that operates from 300 nanometers up to 1,100 nanometers with an option to go up to 1,800 nanometers. It covers the whole horizon and has a vertical sector of 180 degrees. The directional resolution is about 45 degrees. As can be seen by these examples, much work is being done in this field, and a number of detectors are already available from different countries. Future development will certainly continue to proceed rapidly, and detectors will soon be fielded in increasing numbers.

It is possible to give a laser detector system the ability to identify hostile lasers by their wavelength, pulse repetition frequency, and incoming power. It will be possible to judge if the laser is a typical range finder or designator or if the enemy is using LEL weapons. The detectors may be linked to countermeasure devices, such as an armored fighting vehicle's smoke grenade discharger, which would be forced automatically to cover the endangered sector. An automatic system may also be available to decide if the laser threat should be met with counterfire from antilaser laser weapons or from conventional weapons. However, an automatic response triggered by enemy lasers has many drawbacks. The enemy will quickly learn to trigger the protective response with a well-protected laser of its own and then use other laser weapons or conventional means to kill the target or disable the laser detector. In any case, laser detectors will be a necessity on the future battlefield and will be an integrated part of many weapon and intelligence systems. However, they will mainly be of use after the enemy has already fired his laser weapon, although, by then, it may be too late. Automatic responses will be used rarely and even then only in very special situations.

Due to the weight and high costs, it will be impossible to supply individual infantry soldiers with laser alarms other than to place a limited number at the company or battalion level. Even if development efforts succeed, only very sophisticated and, thus, expensive laser detectors will be able to pinpoint the exact position of the hostile laser. It is unlikely that such detectors will be available during the 1990s. Thus, at present, soldiers must wear their protective glasses at all times when protection may be needed. They cannot store them until a laser warning is given, unlike the present situation with gas masks.

CHANGES IN VISUAL BEHAVIOR

The laser weapon is inherently a direct-line-of-sight weapon and should be treated as such. Soldiers on the battlefield will have to keep their heads down and avoid laser beams in the same way that they avoid direct fire. This can be done mainly by using the terrain and other available protection. Anything that is substantial enough to protect soldiers from the effects of conventional guns or rifles will certainly also provide full protection against laser weapons. One problem, however, is that the soldier usually cannot safely see or even hear the laser weapon when it is activated. In combat situations where LEL weapons are used, the soldier may not know when he is being exposed and, thus, will have to stay protected all the time. This is certainly not a possible tactic, as all combat units have a mission to fulfill which requires the soldier to look at the enemy, identify him, fire at him, and sometimes move toward him. All of this requires viewing the enemy either with the naked eye or through some kind of magnifying optics if indirect viewing can be used.

It will be necessary to limit viewing of the enemy to the absolute minimum required when combat units are operating against a laser weapon threat. Soldiers and commanders have to be trained to avoid unnecessary exposure to laser light. The commander has to use as few soldiers as possible to observe enemy movements and positions. The observer has to limit his field of vision to what is strictly required by his task. All soldiers must know that a laser beam may be dangerous even if it comes from a very oblique direction or is only on for a short duration. The laser threat even exists out to 60 to 70 degrees from the optical axis, which means that the peripheral field of view may be as dangerous as the direct field of view. The soldier may limit his field of view by using blinkers similar to those used on racehorses or by observing the terrain through a tube.

Since the use of magnifying optical instruments increases the risk significantly, the soldier has to be advised or ordered not to use magnifying optics any more than absolutely necessary. Such instruments not only magnify the incoming laser light but are also very easy to pinpoint by an enemy laser radar.

A systematic change of visual behavior on the battlefield may reduce the number of eye injuries but will fall far short of avoiding them completely. If the enemy's anti-eye laser weapon is within the militarily important field of vision, the behavioral changes suggested above will be of little or no use at all.

HEL WEAPON PROTECTION

If HEL laser weapons are finally developed and fielded in any considerable numbers, technical and tactical protective measures will also have to be introduced. Such measures will be sought along the same general lines as protective measures against LEL weapons. Since HEL weapons will eventually be designed to shoot down aircraft and missiles by heating the sensitive parts such as canopies, radomes, covers on critical parts, warheads, and, of course, sensors, materials that are laser resistant to as great an extent as possible will have to be developed and used.

Some work on laser-resistant materials has already been accomplished. In the United States, the General Electric Company is developing such a material in its Re-Entry Systems Department. This material is made from multilayered graphite, where each layer acts as a mirror to reflect laser energy instead of absorbing it in the form of heat. The material can be formed in different shapes and in various thicknesses, and it can be mixed with various reinforcement compounds. These materials may also be used to protect space vehicles reentering the atmosphere or to harden military equipment against some of the effects of nuclear weapons. As this material is completely opaque, it cannot be used in goggles, visors, or aircraft canopies.

Several other laboratories are evaluating a material that effectively blocks and diffusely reflects laser radiation over a wide range of wavelengths up to a certain energy level. Above the energy level, the reflective property is lost, all of the energy is absorbed instead of reflected, and the material breaks down. The U.S. Air Force Strategic Air Command is interested in the protection of aircraft fuel tanks, while the U.S. Space Command is interested in the protection of orbital payloads. A number of SDI-related applications are being researched as well. Princeton University's Plasma Physics Laboratory is planning to run high-power tests with a 1-gigawatt CO_2 laser and, possibly, a 20-gigawatt X-ray laser.

Much research remains to be done in order to develop reliable protection against HEL weapons. It may be possible to cope with the thermal effects on material at long distances using laser-resistant coatings and materials, but it will probably be very difficult, if not impossible, to protect laser-sensitive equipment such as sensors, radomes, canopies, and warheads. It will certainly be impossible to protect soldiers in the open at short range.

TACTICAL IMPLICATIONS

There can be no doubt that future battles will be fought in an environment which will contain a combination of conventional weapons, laser weapons, and other laser devices. It will be necessary to adjust present military doctrines and tactics to fit this new situation. Each country will have to formulate a laser philosophy on which its tactics and training can be based. Tactical laser protection methods could either take an active or a passive form. The active methods may include the areas of intelligence, surveillance, counterfire, maneuvers, procedures, and drills. Passive methods against the laser threat may be found in procedures such as cover, concealment, and camouflage. Intelligence and surveillance is not just a question of fielding as many laser detectors as possible. Perhaps the most important procedure is to make all personnel working within this field aware of the facts regarding a laser threat. Everyone responsible for any kind of battlefield threat analysis must have a thorough knowledge of the enemy's laser weapons and laser devices, the enemy's laser philosophy, and how the enemy has used lasers in the past. It is essential that the field commander be given hard facts to be used as factors in his tactical decisions. It must also be possible to use any information collected during the battle about how the enemy laser threat works to improve tactics while the fighting is still going on. This kind of fine tuning will only be really successful if all combat units have already been thoroughly familiarized with the laser threat and know what to do about it before they reach the battlefield.

The most effective tactical countermeasure against any weapon, of course, is to destroy it. The success of counterfire depends upon the availability of very well trained personnel who have often used it against enemy laser positions. Intelligence, surveillance, and rapid information sent along well-established communication links are prerequisites for a well-directed and properly timed counterfire. Since the laser weapons are direct-line-of-sight weapons, it is sometimes impossible for the affected target to shoot back. As a result, counterfire is most likely to be indirect, from artillery, or direct from adjacent weapons that are not targeted by the enemy laser. If the enemy laser position cannot be pinpointed, it will be necessary to cover a bigger area with fire or to use a smoke screen. It has to be remembered that counterfire can only be used after the enemy laser has revealed its position by firing. Thus, it is not possible to stop the enemy laser beforehand. It is only possible to try to destroy the laser weapon while it is in use or after it has been used to stop it from firing again.

Maneuver is of vital importance to all combat units fighting a modern battle. This is true whether there are laser weapons on the battlefield or not. Maneuver is the ultimate tool for achieving surprise and success. The movement of combat units in a laser-infested environment may allow for improved protection of soldiers, equipment, and units. Positions that are highly threatened by hostile lasers should be abandoned if they are not absolutely crucial to the outcome of the battle. When it is necessary to move to a new static position, it should be chosen in an area where the terrain features plants and other obstacles to direct vision and makes it difficult for the enemy to use LEL weapons successfully. When combat units are maneuvering on the battlefield, they must take the laser threat into account. It will always be advantageous to move under cover of rain, fog, dust, or smoke even though such conditions can never offer complete protection. Furthermore, any movements should be covered by terrain and vegetation as far as possible, and it is important to perform the maneuver as quickly and quietly as possible. Whenever feasible, personnel should be moved in vehicles as protection from laser beams. Only drivers and the minimum number of officers and men necessary for command, surveillance, and navigation should be allowed to observe the surroundings. If a driver is hit and blinded, a trained replacement should be prepared to get the vehicle on the move again quickly. This is of utmost importance, since any stationary target on the battlefield has a higher risk of being hit and destroyed by conventional weapons.

Cover, concealment, and camouflage will be effective against laser weapons only if the hostile laser gunner cannot discover or sight his targets or if the cover is substantial enough to stop the laser beam before it reaches its target.

TRAINING IMPLICATIONS

It will be an imperative and urgent necessity to train soldiers and combat units in how to fight on a laser battlefield. So far, there has been general ignorance in most armed forces of the laser threat, and thus there is virtually no awareness of the threat among soldiers and officers. This has mainly been due to the high level of security surrounding the research and development of laser weapons. In countries where this work is being done, it has been decided to keep much of the information highly classified to protect technological secrets and deployment plans. As a result, other countries have not been able to collect sufficient information to plan effective laser weapon protective programs. Now, when laser devices are about to be fielded in great numbers, and the era of laser weapons is emerging, it is certainly urgent to start training individuals and combat units about the laser battlefield environment.

It is not only necessary to prepare soldiers to fight in a laser weapon environment, but the military must also prepare them psychologically. To date, there have been virtually no battle experiences in any country revealing how soldiers would react when subjected to a real laser threat. So far, there are not even any experimental human studies, at least none known to the public, that have addressed this issue, partly because of the risks involved with laser weapons and laser devices. The most basic and difficult question to answer is whether or not an individual soldier will risk his eyes by looking with his naked eye or through an unprotected sight in a direction from which a fellow soldier has been blinded the moment before.

It may even be speculated that a widespread use of laser weapons in a battle will cause a general panic by soldiers forced to fight against lasers. There are few reports and few comments on this problem in the open literature. One report refers to a 1983 secret study on persons who have been victims of laser accidents in the United States by J. A. Wolfe at the Letterman Army Institute of Research. A wide range of reactions were discovered. One scientist, quite familiar with lasers, fainted when he was badly injured, while others who were injured by lasers continued to work. Such a collection of stories is anecdotal at best, and, in the authors's experience, highly suspect, as the injured parties commonly exaggerate the accidental nature of the injury in order not to reveal their often callous disregard of safety requirements.

The reaction to a laser injury appears to depend upon the severity of the injury, but there are also inherent differences depending on the particular individuals involved. One scientist, David C. Decker, gave the following description of his reaction when he was injured by a laser:

> When the beam struck my eye, I heard a distinct popping sound caused by a laser induced explosion at the back of my eye. My vision was obscured almost immediately by streams of blood floating in the vitreous humor and by what appeared to be particulate matter suspended in the vitreous humor. It was like viewing the world through a round fishbowl full of glycerol into which a quart of blood and a handful of black pepper have been partially mixed. There was local pain within a few minutes of the accident, but it did not become excruciating. The most immediate response after such an accident is horror. As a Vietnam war veteran, I have seen several terrible scenes of human carnage, but none affected me more than viewing the world through my blood filled eye. In the aftermath of the accident, I went into shock as is typical in personal injury accidents.

Without any doubt, the facts available are not sufficient to allow any firm conclusions to be drawn. How soldiers will react must be guessed at until substantial numbers of real experiences from battle can be evaluated. However, one way to significantly lessen the risk for panic or other unwanted reactions will certainly be to give the commanders, soldiers, and combat units thorough training about how to cope with a laser threat. One important aim of a good laser training program must be to provide accurate information. In this way, the myths, rumors, and misunderstandings about lasers can be dispelled, and the soldiers can be given a realistic appreciation of what lasers can and cannot do. It is essential to avert beforehand any rumors that can be expected to arise in a battle where laser weapons are used. The second step in a laser training program should be to train officers and men individually on how to use all means of protection available, either of a technical nature such as filters or goggles, or protection provided by the environment in the form of natural terrain features, vegetation, or atmospheric pollution. The overall training of an individual should include instructions on how to react if he himself or someone next to him is hit and blinded by a laser beam. A good training program should include reaction drills which will help the soldier prepare for emergency situations. This will lessen the anxiety and, thus, reduce the risk of a general panic.

Neither proper training of individual soldiers nor fielding of individual eye protectors will eliminate the threat from LEL weapons. It will only be possible to somewhat reduce the risks. The widespread use of laser weapons designed to blind enemy infantry soldiers deliberately and systematically will still lead to mass injuries to human eyes, thus creating a very difficult psychological and medical situation. How this will affect even very well trained soldiers and combat units is an open question.

Flight crews of combat airplanes and helicopters must also be given thorough training. The highest priority will be to teach pilots to avoid the laser threat through low-level flying and taking advantage of any terrain cover, bad weather, and technical aids such as laser alarms and protective visors. Both day and night training for flying with reduced visibility due to eye protective visors must be scheduled and standard procedures established for action when the laser alarm is activated. In the worst case, pilots must be trained in simulators, as it may be impossible to arrange live exercises. When the training of individuals and air crews has been completed, it is necessary to train combat units on how to carry on a combined action on the laser weapon battlefield.

Active tactical laser protection methods involving intelligence, surveillance, counterfire, maneuver, and other procedures must be drilled from the squad level up to the division. Passive methods such as cover, concealment, and camouflage must be learned for protection against both laser weapons and laser radar threats.

CONCLUSIONS

It will be necessary to use many different tactics to meet the laser weapon threat on the future battlefield. By using all available technology and every realistic tactical countermeasure, it may be possible for combat units to carry out their missions and avoid panic and breakdown of the unit. A thorough and professional training of individuals as well as tactical units is a prerequisite if panic is to be avoided and the required military actions carried out. However, there is at present no way to protect dismounted infantry soldiers against LEL weapons. There is no cheap, effective, and acceptable eye protection available, and it seems unlikely that such protective means will become available in the near or even not so near future, as the necessary technology does not seem to exist at present. The side that uses mass-deployed and frequency-agile low-energy laser weapons will have the upper hand for a very long time to come, whether such weapons are used for purposes of attack or defense.

EIGHT

Laser Weapons and International Law

Several governments have voiced their concern about the development and use of lasers as anti-personnel weapons whose principal
effect might result in the permanent blinding of the people attacked. This uneasiness was first made public during a Conference of the International Committee of the Red Cross (ICRC) held
in Geneva in October 1986. That conference passed a resolution
noting that

> Some governments have voiced their concern about the development of
> new weapons technologies the use of which, in certain circumstances,
> could be prohibited under existing international law, appeals to govern-
> ments, with a view to meeting the standards laid down in international
> humanitarian law, to coordinate their efforts to clarify the law in these
> fields and exercise the utmost care in the development of new weapons
> technologies.

In pursuance of this mandate, the International Committee of
the Red Cross decided to hold a round table of experts in June 1989
to clarify the issue. The purpose of the meeting was to consult a
number of experts who could inform the ICRC about the development of antipersonnel laser weapons and their medical and legal
implications. The report from the meeting stated in its final chap-

ter under "follow-up" that the information presented at the meeting needed to be digested and that further research, thought, and discussion were necessary, but the consensus was positive. There will most certainly be future meetings involving scientific and military experts and eventually governments themselves.

This chapter deals with the implications of the possible development of antipersonnel laser weapons on international law. The legal approach taken in this discussion is advanced based on the established international humanitarian law applicable during armed conflicts.

CONVENTIONAL WEAPONS AND INTERNATIONAL LAW

International law has imposed restraints on weaponry for years. Paragraph 1, Article 35 of the 1977 Additional Protocol I to the Geneva Conventions of 1949 states that

> In any armed conflict, the right to the Parties to the conflict to choose methods of means of warfare is not unlimited.

This customary principle that originates from Article 22 of the Hague Rules on Land Warfare dating back to 1907 is fundamental to international humanitarian law applicable in armed conflict. It reflects the mutually balancing principles of necessity and humanity inherent in the law governing the conduct of hostilities. The only measures of warfare that are justified are those which are relevant and proportional to the achievement of a definite military advantage, and they are justified only as long as they are not explicitly prohibited by international law. Those military measures which, from a tactical point of view, are manifestly disproportionate to the human suffering involved are prohibited under international law. This principle can be found in the same Additional Protocol and is supplemented by a somewhat more concrete principle of international customary law stating that

It is prohibited to employ weapons, projectiles, and matériel, and methods of warfare of a nature to cause superfluous injury or unnecessary suffering.

This principle is very old, dating back to the Saint Petersburg (Russia) Declaration of 1868 and the Hague Conferences of 1899 and 1907. Above all, it is related to weapons designed exclusively for antipersonnel purposes, but it also covers weapons designed and produced to fulfill a variety of purposes. Many modern weapons are designed primarily to destroy or neutralize military equipment and matériel. However, most of them may also be used in combat to disable enemy soldiers. It is difficult to decide if the human suffering is needless, superfluous, or disproportionate to the military advantage expected from the use of the weapon. Inevitably, the balancing of suffering against military effectiveness is difficult to quantify, and the process will be a somewhat subjective exercise. If the assessment is focused on a weapon or a method of warfare which is designed exclusively for use against human beings, it may be easier to reach a quantitative conclusion. The Saint Petersburg Declaration, which was adopted by 17 nations in 1868 and today expresses customary law, primarily refers to such a situation. The Declaration states:

That the only legitimate object which States should endeavor to accomplish during war is to weaken the military forces of the enemy: That for this purpose, it is sufficient to disable the greatest possible number of men. That this object would be exceeded by the employment of arms which uselessly aggravate the sufferings of disabled men or render their death inevitable. That the employment of such arms would, therefore, be contrary to the laws of humanity.

In the concluding part of the Declaration, the Parties:

. . . reserve to themselves to come hereafter to an understanding whenever a precise proposition shall be drawn up in view of future improvements which science may effect in the armaments of troops in order to maintain the principles which they have established and to conciliate the necessities of war with the laws of humanity.

These principles were given explicit form in a prohibition in the Saint Petersburg Declaration on the use of highly explosive bullets weighing less than 400 grams. Such bullets had initially been developed by the British to detonate enemy ammunition wagons. Later on, they were modified to explode on contact with a soft target such as a human being. The Russian Government, unwilling to allow another country to take advantage of a bullet that could injure soldiers far more than any earlier projectiles had done, was instrumental in achieving this legally binding agreement prohibiting the use of such bullets under international humanitarian law. By World War I, production of the bullets in question had ceased for all practical purposes. An additional disarmament or arms limitation effect had been achieved *de facto*, albeit not *de jure*.

Another step "to conciliate the necessities of war with the laws of humanity" was taken in 1899 when the first Hague Peace Conference outlawed the use of so-called dumdum bullets. These expanding bullets flattened easily on impact with the human body and made far worse wounds compared to those made by bullets that did not flatten. Dumdum bullets were generally considered to be excessively injurious, and in the spirit of the Saint Petersburg Declaration, the delegates voted 22 to 2 to prohibit their use. Since then, except for a few poorly documented instances, this type of projectile has not been produced or stockpiled for possible use by regular forces. The first Hague Peace Conference, as well as the second one of 1907, also codified a prohibition on the use of poison and poisoned weapons.

The next treaty enacted to prohibit the use of specific weapons was the Geneva Protocol of 1925, which was primarily concerned with nonconventional weapons. The prohibitions in this treaty were not absolute. Reservations were expressed at the time which made it clear that a number of countries considered themselves free to use chemical or biological weapons in retaliation, should their adversaries use them first. This kind of reservation, which differentiates between the first and the second use of a weapon in a conflict, is a concept that is applicable to all humanitarian weapon prohibitions. Since 1925, there has been no comprehensive prohibition regarding the use of any existing weapon category. The total prohibition of projectile fragments which escape detection in the human body by X rays was included in the 1981 United Nations Convention on Prohibitions or Restrictions of Use of Certain Conventional Weapons Which May Be Deemed to Be Excessively Injurious or to Have Indiscriminate Effects. This ban is of limited value, since such a weapon category does not exist and does not seem militarily useful enough to develop. The remaining two regulations from 1981 only established certain restrictions on the use of land mines, booby traps, and incendiary weapons.

The formula for "superfluous injury or unnecessary suffering" is not specific enough to draw any definite conclusions with regard to the legality or illegality of weapons. A specific weapon is not prohibited by the general formula unless an agreement exists between states regarding the weapon in question, where humanitarian imperatives are given precedence over military considerations.

Besides the general principle of "unnecessary suffering," there are two additional principles that must be taken into consideration. First, the "discrimination" principle, which prohibits the use of methods or means of warfare which cannot be directed against a specific military objective and, consequently, are of a nature to strike indiscriminately at both military objectives and civilians or civilian objects without distinction. However, laser weapons cannot be considered indiscriminate. In fact, the opposite is probably true, since laser beams can almost always be directed very precisely against specific targets. The third alternative is the "treachery" or "perfidy" principle. This prohibits certain perfidious uses of weaponry and may also prohibit weapons which are inherently perfidious, although the parameters of this principle are far from certain. In any case, laser weapons will probably not fall afoul of this principle. Therefore, any legal case against the use of laser battlefield weapons cannot rely on the prohibitions against indiscriminate effects, treachery, or perfidy, but only on the prohibition against superfluous injury or unnecessary suffering.

NEW WEAPON TECHNOLOGIES AND INTERNATIONAL LAW

A duty for countries to evaluate new weapon developments and tactics was established in international humanitarian law in 1977. A national screening procedure assessing whether or not the use of new weapons or new tactics under consideration is in line with international legal standards appears to be necessary to all states. Such a procedure is additionally required by the internal regulations of some states such as the former Federal Republic of Germany and the United States. However, these national assessments are not internationally binding. This condition is clearly stated in the relevant Committee Report of the Diplomatic Conference on International Humanitarian Law (1974–1977).

During its 1979 session, the UN Conference on Certain Conventional Weapons (1978–1980) adopted a resolution relating the old dumdum prohibition to the need for caution regarding modern weapon developments. The resolution, dated September 28, 1979, includes the following wording:

> Recalling the agreement embodied in the Hague Declaration of 29 July 1899, to abstain, in international armed conflict, from the use of bullets which expand or flatten easily in the human body. Convinced that it is desirable to establish accurately the wounding effects of current and new generations of small calibre weapon systems including the various parameters that affect the energy transfer and the wounding mechanism of such systems . . .
>
> Appeals to all Governments to exercise the utmost care in the development of small caliber weapon systems, so as to avoid an unnecessary escalation of the injurious effects of such systems

Now that laser weapons are being designed, and laser tactics are closer to being implemented on the battlefield, similar steps of caution should be taken so as to—in the words of the 1979 resolution—"avoid an escalation of injurious effects."

LASER WEAPONS AND INTERNATIONAL LAW

The starting point for an assessment of laser weapons in the international context must be a consideration of the biological effect of these weapons on human beings as compared to the military interests involved. As has already been described in this book, a variety of laser devices and laser weapons can be identified. Laser range finders, designators, and related devices are not designed as weapons in their own right. Any antipersonnel effects of such laser devices are incidental as long as they are used for their designed purposes. When laser weapons are designed for use against materiel such as sensors, optics, canopies, and other equipment, again the effects on personnel are not their principal features. A third category includes laser weapons specifically designed to be used in an antipersonnel mode on the battlefield. Such weapons will be designed to have as large a biological effect as possible, and they will deliberately and systematically be used against human beings. All of these laser weapons and laser devices are capable of causing blindness or severe injuries to the eyes if they are designed with a certain combination of laser wavelengths and other laser properties. Most lasers on the battlefield today and tomorrow will fit into this category. The first two categories of lasers may have an antipersonnel effect on the battlefield that is incidental to their principal use as long as they are only used as designed. The third category will be designed to cause antipersonnel effects. The borderlines between these categories are to some extent blurred. A laser weapon might be designed to blind the electro-optic sensors fitted to a tank as well as the eyes of the tank gunner and tank commander. Another laser weapon designed to be used in an air defense role might affect the sensors on an attacking helicopter as well as the eyes of the pilot. There is even the possibility of designing a laser weapon that could be used for many purposes, thus falling within all three categories.

The use of lasers for range finding, target designating, and similar ancillary purposes is of great military value, and it is difficult to argue that incidental injuries caused by such laser devices are disproportionate to the military interest involved. However, it might be quite another situation form the standpoint of international humanitarian law if a military commander in the field during the conflict or if the staff body of the armed forces before the conflict ordered or made it a part of the tactical doctrines to use these laser devices deliberately and systematically to injure and blind the eyes of the enemy.

The use of laser weapons against military matériel or other equipment can never violate the existing international law. The incidental effects that the antimatériel laser weapons might inflict on personnel have to be tolerated in the same way as is the case with incidental effects on humans from conventional antimatériel weapons. When, for example, an armor-piercing projectile destroys a tank, the wounding or killing of the tank crew is in line with the military interest involved. It is difficult to argue that such incidental injuries are disproportionate to the considerable military value of the attack. Even in cases similar to this involving antitank warfare, the deliberate and systematic use of antipersonnel laser weapons would not contravene the principles of present international humanitarian law.

A more difficult question to address is whether the neutralization of an airplane or a helicopter by blinding the pilot or some other crew member would be considered unlawful. An intentional and irreversible blinding of personnel, as an isolated act, is considered a violation of the principle of suffering. But what if the individual is a pilot in a highly valuable aircraft and constitutes a prime military target? In such a case, the military value of a destroyed enemy aircraft has to be compared to the suffering. This is certainly a difficult procedure, but it seems likely that most experts on international law will arrive at the conclusion that disabling a tank, other armored vehicles, aircraft, or similar fighting equipment by blinding the crew does not contravene the existing law if the laser offers military advantages not matched by other weapons. This view was expressed by C. Greenwood of Cambridge University in England in a presentation before the 1989

ICRC round table of experts. He also stated that the use of laser weapons against personnel who are protected by armor or by their position in a fast-moving aircraft is more readily justifiable than their use against infantry in the open, precisely because alternative weapons are less effective against such protected people.

The remaining question is whether or not the deliberate and systematic use of lasers to blind infantry soldiers or other unprotected personnel should be regarded as unlawful according to existing international humanitarian law. It is necessary to decide whether this question concerns the legality of a specific type of laser weapon or the particular use of a laser weapon or laser device which may also be used in other clearly lawful ways. Some experts would agree that using any kind of technology for antipersonnel purposes that brings about permanent blindness is not in proportion to the legitimate object of warfare. A Swedish expert on international law states that the basic Declaration of Saint Petersburg of 1868 only permits putting the adversary's soldiers out of action, meaning out of action on the battlefield. Although it is permitted to kill combatants under the laws of war and, thus, to put them permanently out of action, it is not permitted to use methods or means of warfare exclusively designed to injure soldiers with the injurious effects lasting, not only for the duration of the conflict, but for the rest of their lives. The same expert claims that in the balance between military interest and humanitarian considerations, an irreversible disablement such as blindness caused by a laser beam must be described as "unnecessary suffering" according to the formulas from Saint Petersburg, The Hague, and Geneva.

Another expert, C. Greenwood, in his presentation to the 1989 ICRC round table of experts very cautiously stated that the discussion of the final position based on the existing law should concentrate upon whether or not the use of laser weapons designed to cause the permanent blindness of unprotected personnel violates the unnecessary suffering principle. According to an article in the U.S. magazine *Army*, there is nothing inherent in a laser, particle beam, or radio frequency weapon which would make its design, development, or use a violation of the laws of war. In the same article, it is stated that blindness is a totally incapacitating injury for a soldier on the modern battlefield; hence, the value of a laser weapon would be great. Furthermore, as blindness and not death is the result, it may be argued that the weapon is actually more humane, as it does not extinguish life although, effectively, it renders enemy personnel "hors de combat." The article even states that, for sole use as antipersonnel weapons, laser and particle beam weaponry are legal under the laws of war. The differing views of experts may be regarded as an indication of the difficulties involved when the existing international humanitarian law is applied. However, there can be no doubt that any person who is suddenly and permanently blinded by a laser weapon will suffer an extremely severe physical and psychological injury. The possible extent of psychological factors was emphasized by D. Warren, a psychologist from California, before the ICRC round table of experts, who stated that he could not see "how we can countenance the prospect of adding to the world's problems by allowing the development and use of antipersonnel laser weapons whose sole purpose is to produce blindness."

AN INTERNATIONAL DOCUMENT ON ANTIPERSONNEL LASER WEAPONS

It is the view of some countries such as Sweden that, irrespective of whether the use of antipersonnel laser weapons should be prohibited under existing international humanitarian law or not, an explicit prohibition against the use of laser weapons designed to cause irreversible injury to the human eye should now be considered for inclusion in an international document.

A specific weapon is not considered prohibited under international law unless there is a treaty to that effect. Bearing in mind the wide range of both civilian and military uses of laser devices, a total ban on all use of lasers and laser weapons would be unrealistic. However, there are strong reasons for seeking to establish a formal prohibition in the area of antipersonnel use.

A prohibition could be achieved either through focusing on the methods of warfare or through focusing on the means of warfare. In the first case, there would be an explicit ban on the deliberate and systematic antipersonnel use of any kind of lasers that would cause superfluous injury or unnecessary suffering. This would be a prohibition of certain tactics or methods of warfare. In the second case, which could be called the hardware approach, the use of antipersonnel laser weapons specifically designed to blind or injure human eyes would be banned. The two approaches could also be combined in the same document. Such a document might be drafted as follows:

1. Confirm the accepted legal prohibition of certain methods of warfare which can be related to laser technology.
2. Introduce a prohibition on the use of certain antipersonnel laser weapons.

Such a dual ban on tactics and weapons would, of course, be the ideal solution from the control point of view. However, it must be recognized that the only control that seems possible would be *post facto* on the battlefield. However, this is far from useless. Any systematic antipersonnel use of lasers would soon be revealed, and, at least in armed conflicts of a limited nature, the political costs to the user involved could be substantial. There can be no doubt that the introduction of a great number of anti-eye laser weapons in the armed forces would add a very severe threat to the majority of soldiers on the battlefield. A document protecting human beings on the battlefield from being deliberately blinded for the rest of their lives would be an important new element in the tradition of humanitarian law.

The laser incidents that have occurred between the U.S. and Soviet ships and aircraft resulting in the temporary blinding of military personnel mean that even peacetime laser threats have forced the superpowers to react. The governments of the United States and the Soviet Union have realized the need for an international regulation to serve as a preventive measure against incidents between their armed forces and ultimately to protect their own personnel. In 1989, a bilateral agreement on the prevention of dangerous military activities was signed in Moscow. The second article of this agreement stipulates that each party shall take necessary measures directed toward preventing dangerous military activities, including "using a laser in such a manner that its radiation could cause harm to personnel or damage to equipment of the armed forces of the other Party." However, this agreement, which became effective on January 1, 1990, is limited to activities "in proximity to personnel and equipment of the armed forces of the other Party" during peacetime. Even so, the agreement must be considered a very strong indication that the big powers are taking the laser threat seriously, and it is also an indication of the strong interest in the development of laser weapons. The agreement is an important step forward, but it must be widened to include all countries as well as to conform to the international humanitarian laws applicable in armed conflicts.

NINE

Conclusions and Consequences

Laser weapons are undoubtedly here to stay. A few low-energy laser (LEL) weapons are already a fact, and more will come as the technological developments continue. In addition to this, other low-energy laser devices such as range finders and designators may deliberately be used as weapons. High-energy laser (HEL) weapons are not only different in concept but are also far from being realized. There may eventually be a few HEL weapons fielded some years into the next century. This means that HEL weapons do not present immediate operational and training problems to military staffs and units.

The introduction of LEL weapons on the battlefield for various purposes will have many consequences. Some of these will be comparatively far-reaching and will not only affect combat units but also might possibly alter the social structure of the combatant nations as a result of the numbers of casualties and disabled veterans. There will certainly be military consequences of many kinds, serious medical ramifications, and prewar changes in the laser industry, as well as postwar consequences for society in general. Laser weapons will probably even effect the development of an international law. The aim of this chapter is to summarize some of these major consequences.

215

MILITARY CONSEQUENCES

Low-energy laser weapons will soon be deployed and used within all branches of the military service in every country, although the numbers and levels of sophistication will vary. Most LEL weapons will be deployed within the ground forces, and these will be less sophisticated than those designed for use in air and sea applications. This also means that the most expensive LEL weapons will be found in ground-to-air and ship-to-air applications. Almost all LEL weapons will be fielded within traditional combat units and used alongside conventional weapons, since this combination will produce the largest possible effect.

Battlefield laser weapons are tactical rather than strategic weapons. They will affect operations performed by large combined forces only insofar as the actual fighting on the battlefield between smaller units is affected. If large-scale and skillful use of battlefield laser weapons significantly tilts the battle in favor of one of the adversaries, this will obviously have some sort of impact on the overall operations.

Thus, laser weapons will have most of their effects on the battlefield at the small combat unit level, such as companies and battalions, and will mostly be operated by individual soldiers. The weapons will be used to kill or blind sensors or other laser-sensitive devices. It cannot be ruled out, however, that LEL weapons might possibly be used on a large scale in an antipersonnel application for destroying or blinding eyes.

All combat units will have to learn how to use and benefit from their own laser devices such as range finders and designators. This will require a thorough and realistic training course, which is, to some degree, difficult to achieve due to the hazards involved with the use of many lasers. As a result, new laser devices that are safe to the eye may be developed and delivered in increasing numbers. The problem is that lasers that are safe to the eye are more expensive, and a lot of money has already been invested in the retinal hazard lasers that have already been operationally deployed. It can also be argued that there is no big difference between firing with a hazardous laser beam and firing with live munitions. This may be so in combat, but in training it is a military requirement for most armed forces to be able to use their laser range finders in two-sided exercises, where blank munitions are used.

Training with laser weapons deployed in the combat units of one's own forces has to be carefully planned in order for the commanders and men to learn how to take full advantage of all of the laser's power. The use of laser weapons against enemy sensors or personnel behind sights, electro-optical equipment, or vision blocks requires a high degree of knowledge about the target itself. It is necessary for the operator to know if the laser weapon being used is effective against the intended target, which, in turn, depends on the sensitivity of the target to specific laser weapon properties—wavelength, pulse length, and pulse energy. The operator should also know the level of protection that the target has against hostile laser beams. Units using laser weapons should not be allowed to expose or endanger themselves and their weapons unless they are certain that they have chosen a proper laser target. This is especially important with laser weapons, since it is very difficult to determine what the effect on the target actually is once the laser has been used against it, the effect being very different from that of a conventional weapon—blinding as opposed to the explosion resulting from the impact of a high-speed missile.

The introduction and use of battlefield laser weapons not only means that the military must implement a complete laser weapon philosophy and devise new training methods, but they also must ensure that all possible threats from enemy lasers have been identified and properly addressed before an attack takes place.

THE SOLDIER AND THE LASER THREAT

The existence of hostile laser weapons on the battlefield will undoubtedly pose several problems to the individual soldier.

Soldiers must face the risk of getting their eyes injured or destroyed, especially when they are deliberately observing the enemy through optical devices. After they have spent some time on the battlefield and observed what happens to their fellow soldiers, they will be well aware that the enemy may use laser radar which can reach their own optical devices and send very agile, eye-threatening beams which will injure or destroy their own vision. Watching their friends go blind will create tremendous psychological problems. As a result, many individual soldiers will deliberately avoid looking in the direction of the enemy, which, of course, will severely reduce their combat efficiency. This will certainly affect the performance of their combat units and may mean that, when the outcome of an ongoing battle is close, the battle will be lost.

The situation will be even worse if the use of anti-eye lasers threatens not only the soldiers looking through optical devices but anyone else looking toward the enemy with the naked eye. For infantry soldiers fighting in close combat, looking at the enemy is a must. A high risk of getting blinded will certainly mean that the soldier will not look as carefully or as much as is really needed, which means that combat efficiency will be markedly reduced.

Some may argue that laser weapons are not more dangerous to combat soldiers than bullets, shrapnel, and land mines. This is certainly true, and it may very well be that the overall danger from anti-eye laser weapons is less. However, the psychological trauma associated with the risk of blindness will have a more severe impact on the psyche of individual soldiers. It is possible for a soldier to accept the risk of being wounded by a bullet or shrapnel. These wounds usually heal, and, even if not, he can get along even with the loss of a limb. It is also possible for most soldiers to accept the risk of being killed. However, it is quite another thing to lose one's eyesight. Human beings instinctively protect their eyes more than any other part of the body. If the soldier thinks during combat that he is facing a high risk of becoming blinded for the rest of his life, this could create a more severe psychological problem on the battlefield than any injuries caused by conventional munitions.

One way to eliminate this psychological fear completely would be to supply the soldiers with efficient protection against the laser weapon threat. However, for the time being, this is not possible without degrading vision in an unacceptable way. The eyes can be protected against a few laser devices but not against the frequency-agile anti-eye laser weapons. A lot more research has to be performed in this field, and every possibility needs to be investigated. Some work is being done, but the prospect of finding a cheap, efficient, transparent, and comfortable eye protection that will solve all of the major problems seems slim.

To cope with the laser weapon threat and with the accompanying psychological symptoms, the individual soldier must be very carefully trained. In order to allay the soldier's fear as much as possible, he must know exactly what he is up against and be very well informed about what the laser threat actually looks like. Knowing what lasers the enemy is using and how to identify them is basic and necessary knowledge. The soldier should also know as much about the enemy's laser weapon philosophy as possible, which basically means that he has a thorough knowledge of how and when hostile lasers may be used against him. Some of these facts may not be fully understood until the battle has been fought for some time. In this situation, it is of crucial importance that all experiences of enemy use of lasers be spread rapidly among the soldiers. Knowledge is a very basic tool that can be used to fight fear.

The soldier should be trained to use the most sophisticated visual instruments without hesitation when facing an enemy equipped with laser weapons. He must be able to perform all of his assigned tasks using as little time and as small a visual field as possible while he is viewing enemy targets to reduce the risks of eye injury. He has to be able to avoid unnecessary exposure to hostile laser beams. The use of technical aids such as laser protective goggles, filters of different kinds, and laser detectors must be part of his training, and it is mandatory that he knows the performance and various limitations of all such technology.

It is necessary that all soldiers know how to take care of and help blinded fellow soldiers. Victims have to be brought out of the firing line and taken into cover. They need to be calmed down and handled in a way that prevents panic.

As always, the key personnel on the battlefield are the commanders at the battalion, company, and platoon levels. Their training to cope with the laser threat posed to their units must be extensive and in line with what is expected from them when they are leading their men through a laser weapon environment. They have to be familiar with the basic technology behind laser weapons in order to have a thorough understanding of what is happening or what might happen on the battlefield. They have to give the best possible orders and instructions to their soldiers and be prepared to cope with new and so far unknown enemy laser weapons, tactics, and philosophy. In the end, their actions will make the difference between success and mass panic.

Thus, in order to successfully meet the threat from anti-eye laser weapons, all armed forces must prepare new training manuals and training programs.

THE AIRCRAFT PILOT

In battle, there are many reasons to single out the aircraft pilot as the highest priority target for low-energy laser weapons. When a pilot is hit by a blinding laser beam at the final stage of his attack, he is in real trouble. The laser beam may blind him for less than half a minute. However, that is more than enough time to force him to leave the aircraft by parachute or crash. For a helicopter pilot as opposed to an airplane pilot, the situation is much worse for he has no alternative but to crash, since he has neither an ejection seat nor a parachute.

A pilot's eyesight is very difficult to protect, since his field of vision cannot be degraded in any way. For the time being, it is not possible to design, for instance, protective visors without blocking too much of the light. Since the pilot must look directly at the target or at the hostile weapon sites around the target while he is attacking, restricting his field of vision is not a feasible solution either.

The laser threat to the pilot can be reduced through training techniques that teach avoidance of the laser beam for as long as possible, and then what to do following an eye injury. The techno-logical development of sights and sensors in new aircraft or helicopters should concentrate on indirect methods of viewing, at least for the final phase of a direct attack, where the laser threat will be the greatest. If present-day aircraft or helicopters are to be used for the next ten years or even more for direct attacks, they should be retrofitted with a means of indirect viewing. These solutions will certainly be very expensive, but they are necessary in order to protect these very expensive weapon systems and their highly skilled pilots.

EFFECTS ON SENSORS

One of the main reasons for developing laser weapons is to kill or block sensors. This threat has to be countered when the sensors are first designed, and there will be a constant running battle between countermeasures and counter-countermeasures. However, it will still be much easier to handle the laser threat to sensors than the threat to human eyes. The simplest way may be to build in an easily replaceable part that will break when the laser beam hits, thus protecting the key parts of the sensor. Sensors may also be designed with a limited visual field for use in only a small part of the spectrum.

The relationship between the high initial cost for most of the high-technology sensors and the rather lower cost for protective measures will be such as to promote protection. This will be the case especially when new sensors are designed, since it will be difficult and expensive to install protection for existing sensors in a retrofit program. For some very common sensors, such as image intensifiers, any protection by a filter is almost out of the question since they are required to be extremely sensitive to light, and the filter will take away that sensitivity.

This means that all soldiers, sailors, pilots, and commanders who operate sensors must both be aware of the laser threat and be thoroughly trained in the proper procedures to follow for the operation and protection of their particular system. It goes without saying that the development of protective measures must proceed with a high priority.

If high-energy laser weapons materialize at some point in the beginning of the next century, then sensors must be heavily protected to be able to cope with the severity of such a threat. Otherwise, most of them will be burned out or the glass in front of them crazed.

COMBAT UNITS

It is not only the individual commanders and their men who will have to modify their activities on a laser-infested battlefield. The combat unit as a whole will also certainly have to change its tactics. Combat units will have to be deployed on the battlefield with the laser threat constantly in mind. The cover offered by the terrain has to be used in the best possible way to avoid unnecessary exposure of the soldiers. Only a limited number of soldiers and sensors should be used for surveillance of the battlefield.

Both sensors and human observers have to be kept in reserve as replacements for those hit by hostile laser beams. The time spent on the attack or in defensive actions or for observation in direct visual contact with the anti-eye and anti-sensor laser weapons of the enemy should be kept to an absolute minimum. When direct contact is necessary, the equipment that is most laser damage resistant should be used as long as possible before more fragile systems are used or before human eyes are put at risk.

All combat units must be prepared for laser eye casualties. The medical staff should be trained and organized to handle such a situation. The medical facilities to the rear of the divisions and corps must include eye hospitals with a sufficient number of specialists to be able to manage a relatively large number of soldiers with laser-induced eye injuries within a short period of time.

MEDICAL CONSEQUENCES

The number of eye injuries caused by conventional battlefield weapons during war has been limited in the past compared with other injuries. However, even this small number has increased from about 1% of the total a hundred years ago up to 10% in recent conflicts. The actual figures for World War II and the Korean war were the same, 2%, whereas those for the Six Day war in the Middle East in 1967 and the 1973 Yom Kippur conflict were 5.6% and 6.7%, respectively. This increase in the number of eye injuries has not been due to the introduction of lasers but rather is the result of the larger number of very small splinters and fragments released by modern conventional weapons.

If anti-eye laser weapons are deployed and used on the battlefield in large numbers, there will undoubtedly be an increase in the number of eye casualties. The exact numbers are impossible to predict, but if the infantry is supplied with these weapons and uses them in combat, there could be more than three to four times as many eye injuries as were registered during the latest conflict. Even if anti-eye laser weapons are not deployed, an increase in eye injuries caused by other laser devices and antisensor laser weapons can be expected.

A complete plan must be worked out detailing how to handle eye casualties from the very moment the beam damages the eye down the medical line to the final and long-term treatment. Some of the injuries will only be treatable by a highly trained ophthalmic surgeon. Speed is a crucial factor, since the patient should undergo surgery no later than two or three days after the injury, with the surgery performed by highly skilled eye surgeons at a hospital with extremely clean and sophisticated operating facilities. Only a few hospitals capable of providing the required treatment exist in peacetime society, even in industrialized countries. The number of eye doctors with the necessary training, surgical background, and operating facilities is far from sufficient. More doctors have to be trained, and more eye hospitals have to be built if we are going to be able to handle the consequences of the laser battlefield properly. It will certainly be a very long term investment to begin training doctors and to establish new and expensive facilities that will not really be necessary during peacetime. Few countries, if any, will go to these lengths, which means that most of the injured soldiers will not get the best possible treatment. It can be expected that many soldiers will not get any treatment at all, and, even if a small chance existed to save some vision, it will be tragically lost.

Even in cases where proper medical treatment is administered, some of the injured soldiers will be permanently blinded for the rest of their lives due to the very severe nature of their injuries. Very early in the chain of medical treatment, it is necessary to include careful selection of those casualties who can have a considerable part of their vision restored by available treatment. The limited medical resources during wartime will give the hopeless cases a low priority. The fact that it will be very costly to build up even a limited capability to handle eye casualties is one of the reasons for the tremendous cost-effectiveness of relatively cheap anti-eye laser weapons.

A future scenario where most soldiers injured by anti-eye laser weapons receive no better treatment than a white cane is terrifying but not unlikely.

CONSEQUENCES TO THE LASER INDUSTRY

Acquisition of laser weapons will certainly lead to a great expansion in that part of the electronics industry that handles military laser contracts. The industry has to take care of some of the research and development as well as all of the production of these new weapons. This will in turn lead to the development of a more advanced laser technology handled by the same industry.

The field of research and development will undoubtedly grow, and much of it will be devoted to creating countermeasures and counter-countermeasures in an increasing competition for better means, methods, and measures.

In a world in which defense expenditures are decreasing, money will be directed from the research, development, and acquisition of conventional weapons to new weapon families such as laser weapons due to their increased cost-effectiveness. Therefore, it may be concluded that another burgeoning area that will receive increasing financial support in the future is the technology for protection against hostile laser beams. If the problem of protection can be resolved in a cheap and effective way that will make these protective devices easily available, the development and fielding of anti-eye laser weapons will not be as interesting from a military standpoint as it is now.

Based on the trend of current research, the growth of the current laser industry will ultimately create a large and possibly enormous market within the laser weapon field in the future.

EFFECTS ON SOCIETY: IMMEDIATE AND POSTWAR CONSEQUENCES

Blindness is certainly a very serious type of handicap for any person. Even a serious degradation of vision, resulting in, for instance, an inability to read rapidly or drive, will in the end be almost identical to becoming totally blind. For a young person such as a soldier, blindness is a devastating blow that would shatter all plans and hope for the future.

An individual's self-confidence is significantly lowered or lost completely as a result of sudden severe visual impairment, which is compounded by dependency upon the people who must care for him. Blindness is more devastating to a young person than the loss of a limb or any other severe permanent injury. Even if a very aggressive person tries hard to get along with a visual disablement, in most cases he cannot live a full and normal life. Rehabilitation is long and arduous.

The family taking care of the blind soldier will have a heavy and long-term burden to carry. At least in the beginning, the soldier will be highly dependent upon them. His parents or his wife may have to manage for a long time without the support of a healthy person. Instead, they will have to take care of the blinded soldier until he is rehabilitated, which may be for the rest of his life.

Wartime and postwar society will have to care for thousands of blind men in the aftermath of a battle fought with lasers. Family support and voluntary support from organizations and individuals will not be enough. Society at large will have to commit huge financial resources to help all of these blind victims by investing in different kinds of assistance programs. This will include programs to provide living accommodations adjusted to fit the needs of the blind, transportation, suitable jobs, recreations, and education designed for blind people.

Without minimizing the seriousness of a severe visual disability, the emphasis of the preceding discussion may give too gloomy a picture. People do cope with blindness, and younger people are usually more adaptable than others. Also, the very laser technology that will have produced the problem may also be involved in mitigating it. Unquestionably, more funds and resources will have to be made available for the research and development of aids for the blind including laser canes for mobility, miniature laser product code readers for stores to aid in shopping, better aids for low-vision reading, and perhaps even automatically controlled automobiles. This program may even aid an ever growing segment of today's society who are relatively neglected now, such as those who are blind or severely visually disabled as the result of trauma or disease. One of the leading ocular problems at present is an "age-related" disease formerly termed "senile" macular degeneration, which involves subretinal hemorrhages similar to those following a laser injury, with a similar loss of vision. Possibly, the surgical therapy and rehabilitation training development for battle casualties may assist these people as well. Any research or advances that are made to assist the military casualties will have an immediate and important effect on civilians. Thus, in some way there could be some benefit to society from the military involvement with blindness to offset the unquestionably great burden imposed on the country by the large number of visual casualties that will follow hostile military activities.

The difficulties that any postwar society will have to face under these circumstances should not be underestimated. If possible, we should consider avoiding these problems altogether by banning anti-eye laser weapons from the battlefield under international law. The staggering consequences for the individual, the family, and society are not only severe but also extremely long-term. The changes will last a lifetime.

INTERNATIONAL LAW

The question of the use of antipersonnel laser weapons has already raised international concern. The matter is now before international bodies such as the United Nations and the International Committee of the Red Cross (ICRC).

It has been pointed out by some experts on international law that the deliberate use of anti-eye laser weapons is in direct conflict with existing international law on unnecessary suffering. This viewpoint is not undisputed. Some experts believe that such weapons can only be prohibited if the international law is made more specific for laser and other blinding weapons.

The introduction of anti-eye laser weapons on the battlefield will certainly add a completely new weapons concept to international law. Furthermore, neglecting legal considerations for a moment, it may be questioned from a military point of view whether it is necessary to escalate the conventional battle scenario by introducing the mass deployment of anti-eye laser weapons. If such weapons are fielded, they will most certainly be available to both sides in the conflict. There is nothing about this technology that prevents almost any nation on earth from buying these weapons or developing them and adding them to its present armory. This means that the staff officers will have to take into account not only their own benefit from these weapons but also the threat that their own troops will face from hostile, frequency-agile enemy anti-eye laser weapons. Are the military advantages worth the risks of a potential strategic as well as medical nightmare?

Prohibition of the systematic use of anti-eye laser weapons would be a viable option. Such a prohibition must be carefully designed. It is out of the question to ban laser weapons or laser devices altogether. The development of antisensor and anti-matériel laser weapons and other valuable laser devices must be allowed, as must their procurement and use on the battlefield as long as they are not deliberately and systematically used against human eyes. Blinding electronic sensors and eyes is sometimes the same thing, which somewhat blurs the borderline between antimatériel and antipersonnel laser weapons. Possibly, the prohibition should specify the deliberate and systematic use of all lasers against human eyes. Thus, incidental blinding when lasers are used for their assigned tasks cannot be deemed unlawful. Furthermore, the use of lasers to blind sensors on attacking enemy aircraft, helicopters, and armored vehicles cannot be deemed unlawful even if the final destruction of the target is caused by blinding the crew. Such weapons should be considered anti-matériel, as the main purpose of their use is to destroy the matériel target, that is, the aircraft.

The most terrible weapon on the future battlefield may be the small, cheap, and, thus, mass-produced anti-eye laser weapon which will have the power to cause mass blindness among infantry soldiers. This will result in devastating postwar problems.

The laser technology to support the military development of anti-eye laser weapons is already here and is moving rapidly ahead, not only within military programs but increasingly within business and industry. That means that what is possible today will be even more readily achieved in the future. Most nations will be able to mass-produce cheap and effective anti-eye laser weapons in the future. As this scenario materializes, battlefields will become different. The Persian Gulf war showed very clearly to the public that the use of lasers for guidance has already changed the battlefield. The future use of lasers as weapons will change it even more. Low-energy antipersonnel weapons that are mass-deployed will have an impact on every commander making decisions and every soldier fighting in battles in which these weapons are used. There will surely be many armed conflicts during this and following decades where antipersonnel laser weapons could be used. Current political developments all over the world may well lead to a long period of unrest when a number of new small nations are trying to establish their borders. Economic interests, classical enmities, and strong feelings of nationalism will cause a number of limited armed conflicts, border clashes, and even small wars, as is already the case in parts of Europe.

These small-scale conflicts will certainly not be of the same character as the high-technology Persian Gulf war. Most of these, not only in Europe but all over the world, will be dominated by conventional low-technology weapons mingled with a few high-technology systems. Small, cheap anti-eye laser weapons may very well be used, along with conventional devices such as range finders and target designators deliberately and systematically directed against unprotected eyes, causing a great number of eye casualties.

APPENDIX I

Recommended Readings

A detailed presentation of laser safety can be found in several books, of which the most complete may be *Safety with Lasers and Other Optical Sources* by D. Sliney and M. Wolbarsht, published in 1980 by Plenum Press, New York.

This book is based entirely on public material. To a large extent, information on current military laser developments has been obtained from the following sources:

Air Force Magazine
Armada International
Armed Forces Journal
Armed Forces Journal International
Army
Aviation Week
Aviation Week & Space Technology
Defense
Flight International
Guardian
Health Physics
J. Hecht, *Beam Weapons*, Plenum Publishing Corp., New York, 1984.
J. Hecht, *Understanding Lasers*, Howard W. Sams & Co., Indianapolis, 1988.

IEEE Spectrum
Independent
International Defense Review
Jane's Defense Weekly
Journal of Electronic Defense
Journal of Peace Research
Laser Focus
Lasers & Optronics
Letterman Army Institute of Research, Proceedings of Conference on Combat Ocular Problems, San Francisco, October 1980.
Military Electronics/Countermeasures
Military Review
Military Technology
NATO'S SIXTEEN NATIONS
New Scientist
Nordic Journal of International Law
Soldat und Technik
Soviet Military Power, U.S. Government Printing Office, Washington, D.C., March 1987
Sunday Telegraph (London)
The Sunday Times
U.S. Army Technical Bulletin Med. 524, Control of Hazards to Health from Laser Radiation, Washington, D.C., June 1985.
Washington Times

APPENDIX II

Metric–English Systems Comparisons

The units familiar to Americans as part of the "metric system"—gram, liter, bar and millibar, calorie and kilocalorie—are for the most part not used in scientific work anymore.

The "metric system" has a single fundamental unit for any measurable quantity. Multiples of the units, both larger and smaller, are formed by the use of decimal prefixes. The most often used prefixes are those multiples which differ by a factor of 1,000 from each other.

The metric units used in this book are:

length	meter
mass	kilogram
volume	stere
power	watt
energy	joule

The decimal prefixes are listed below. It should be noted that the use of the hecto, deka, deci, and centi prefixes is discouraged.

Prefix	Symbol	Multiplication factors
exa	E	1 000 000 000 000 000 000 $= 10^{18}$
peta	P	1 000 000 000 000 000 $= 10^{15}$
tera	T	1 000 000 000 000 $= 10^{12}$
giga	G	1 000 000 000 $= 10^{9}$
mega	M	1 000 000 $= 10^{6}$
kilo	k	1 000 $= 10^{3}$
hecto	h	100 $= 10^{2}$
deka	da	10 $= 10^{1}$
deci	d	0.1 $= 10^{-1}$
centi	c	0.01 $= 10^{-2}$
milli	m	0.001 $= 10^{-3}$
micro	μ	0.000 001 $= 10^{-6}$
nano	n	0.000 000 001 $= 10^{-9}$
pico	p	0.000 000 000 001 $= 10^{-12}$
femto	f	0.000 000 000 000 001 $= 10^{-15}$
atto	a	0.000 000 000 000 000 001 $= 10^{-18}$

The metric units are approximately equivalent to units in the "English system" as follows:

Metric	English
1 meter (m)	39.37 inches
25.37 millimeter (mm)	1 inch
1 micrometer (μm) [formerly micron (μ)]	3.937×10^{-5} (0.00003937) inch or 39.37 microinches
1 kilometer (km)	0.62137 mile $= 1,093.6$ yards
1.60935 kilometer (km)	1 mile
1 square centimeter (cm^2)	0.155 square inch
6.45 square centimeter (cm^2)	1 square inch
1 kilogram (kg)	2.2046 pounds $= 35.274$ ounces
0.4536 kilogram (kg)	1 pound $= 16$ ounces
1 watt (W)	0.00136 horsepower
746 watts (W)	1 horsepower

1 joule (J) = 0.239 calorie

9.48×10^{-4} (0.000948) British thermal unit (BTU)

1,054.8 joules (J)

1 British thermal unit (BTU)

The calorie familiar to most Americans for food values is the kilocalorie or large calorie, equivalent to 4,186 joules.

The wavelengths of some common military lasers in both systems follow:

	Metric (nanometers)	English (microinches)
Argon	488, 514.5	19.21, 20.26
Ruby	694.3	27.33
Neodymium:YAG (Nd:YAG)	1,064	41.89
CO_2	10,600	417.32

Index

237